D0344428

LAND

ON

FIRE

LAND

THE NEW REALITY OF WILDFIRE IN THE WEST

GARY FERGUSON

TIMBER PRESS · PORTLAND, OREGON

FRONTISPIECE **Throughout the West, the first fifteen years of the new millennium saw tremendous growth in the size and force of wildfires. In New Mexico, for example, the Cerro Grande Fire of 2000 destroyed a record 280 homes, and four years later, 242 homes were lost in the Little Bear Fire, shown here, ignited by a lightning strike in the Lincoln National Forest.**

Photo and illustration credits appear on pages 202–205.

Published in 2017 by Timber Press, Inc.
The Haseltine Building
133 S.W. Second Avenue, Suite 450
Portland, Oregon 97204-3527
timberpress.com

Printed in China
Jacket and text design by Patrick Barber

Library of Congress Cataloging-in-Publication Data

Names: Ferguson, Gary, 1956- , author.
Title: Land on fire: the new reality of wildfire in the West / Gary Ferguson.
Description: Portland, Oregon: Timber Press, 2017. | Includes bibliographical references and index.
Identifiers: LCCN 2016045506 (print) | LCCN 2016054555 (ebook) | ISBN 9781604697001 (hardcover) | ISBN 9781604698121 (e-book)
Subjects: LCSH: Wildfires—West (U.S.) | Wildfires—West (U.S.)—Prevention and control.
Classification: LCC SD421.32.W47 F47 2017 (print) | LCC SD421.32.W47 (ebook) | DDC 634.9/6180978—dc23
LC record available at https://lccn.loc.gov/2016045506

CONTENTS

PROLOGUE

ACROSS MY THIRTY-FIVE YEARS OF WRITING ABOUT THE NATURAL and cultural history of the American West—logging nearly a quarter million miles of highway and some 30,000 miles of trail along the way—wildfire has been a common companion. I've seen its handiwork all around my longtime home in southern Montana; how it's driven new generations of aspen and lodgepole forests in the Beartooth Mountains; and farther to the south, in the outback of Yellowstone, the way it's cleaned and pruned the Douglas-fir and Engelmann spruce forests. Flames at my back have sent me scurrying like a startled mouse out of the lonely folds of Hells Canyon, while big burns have eaten beyond recognition some of the landscapes I roamed in my youth: slices of the Sawtooth Mountains of Idaho, the Weminuche Wilderness of Colorado, the southern uplands of Utah.

When I first came to the West as a young man in the late 1970s, wildfire was still seen largely as a destructive force, which of course at times it can still seem today. But across the decades I've also come to know it as a powerful agent of healing, a mighty wand that wipes the land free of disease and insects and fallen timber to create a stage for healthy, altogether magnificent new flushes of life. By returning essential nutrients to the soil, fire allows a flush of grasses that can provide especially nutritious graze for elk and bison, not to mention food for dozens of species of ground-foraging birds. At the same time, small mammals who feed on the seeds of those grasses tend to increase in number after a burn, in turn providing food for hawks, owls, coyotes, and the like.

Lately, though, I've also been witness to this land changing, increasingly being wrung dry by severe episodes of drought. And as a consequence, wildfire is establishing itself as a far bigger, much more forceful presence than ever before. In many recent years my neighbors and I have choked on smoke from burning forests, have turned our heads up to the August sky looking for rain until our necks hurt, and on several occasions have packed up a few precious belongings and evacuated our homes, hearts in our throats, in the face of advancing flames. Despite the hubris humans have so often brought to our relationships with the natural world (in the case of wildfire, once believing we could all but eliminate it), fire has proven awfully good at dealing blows to swagger.

What will we do as tens of thousands of acres of conifers, stressed by drought, succumb to infestations of beetles and disease, creating fuel loads that sooner or later will feed massive infernos? How do we control the invasive grasses and shrubs flaring across western landscapes, not only diminishing grazing values but also serving as flash fuses for the rapid spread of wildfire? With annual costs of fire suppression already in the billions, how do we fund not only future firefighting but also the prescription burning and forest-thinning operations needed to reduce the risk of major conflagrations? And even if we do find money for things like prescribed burns, will communities allow them, given growing concerns about air pollution as well as the possibility (though small) that such burns can now and then get out of control? And finally, how will the astonishing webs of life that are now strung across these great landscapes—encompassing salamanders and grizzlies, pikas and pinyons—be changed by the conditions that today allow wildfire such a heavy hand?

Like it or not, today seventy-five million people find themselves living in the western United States in a time of fire. And fire—like other big forces of nature—doesn't suffer fools. It has no patience for our stubborn refusals to acknowledge the realities of our time. If we expect to minimize loss and suffering in the decades to come, we need to start making some serious changes to get along better with wildfire, not to mention living in ways that minimize the climate shifts that are making fire an ever more dangerous force.

Maybe the first step is simply to ask questions. To learn—from the men and women whose lives turn around wildfire, as well as from the land itself. To educate ourselves toward some deeper understanding of how to live intelligently, even gracefully, in what has clearly become a land of flames.

131

LIVING FIRE

Not your grandparents' landscape

AS MORNING ROLLS ACROSS A SLICE OF PONDEROSA AND DOUGLAS-fir forest in southern Colorado one August day, thin sheets of clouds are dissolving into tatters and then scattering in the wind. More clouds come later, in the heat of the afternoon, some with dark curtains hanging from their bellies, telltale signs of rain; but the air is so warm and dry that no drops reach the ground. Then in midafternoon, a bolt of lightning blasts out of the sky, discharging at the top of a 60-foot ponderosa pine snag on a steep south-facing slope. No one is around to see the hit, the brilliant flash of the lightning bolt carving a channel through the air with such force that when the channel fills again, it creates an ear-shattering boom of thunder.

The dry wood explodes into the surrounding forest, sending a light spray of sparks into the air. As the base of the snag smolders, a thatch of dry fescue grass, strewn with bits and pieces of fallen limbs, begins to burn. Fanned by a gusty wind—common to dry lightning events—the flames start to spread. Embers begin to pop and blow. One lands in the shallows of a small creek, fizzling with a soft hiss. Another falls on a slab of granite and burns itself out. But still another flies about 20 feet to land in a patch of cheatgrass and small branches, the latter blown off trees killed in a pine bark beetle infestation.

Thus begins another wildfire in the American West.

IN THE NOT-SO-DISTANT PAST, WHEN WILDFIRES BURNED UNRE-strained, they played an essential role in shaping the forests. To better comprehend this role, it might help to take a look at one specific kind of forest community, found not only in southern Colorado but also from New Mexico to Montana, west through Idaho and Utah and parts of Nevada to Oregon, Washington, and California. The centerpiece of this community is the ponderosa pine. The ponderosa's signature features—beautiful cinnamon-colored bark, which on mature trees is thick and furrowed, as well as the tree's ability to self-prune, which means dropping its lower branches when they no longer have access to the sun—are in fact evolutionary adaptations linked to wildfire. The tree's thick bark allows it to withstand flames without suffering damage to the cambium layer (the part of the tree that produces both new bark and wood), while the self-pruning of lower branches keeps ground fires from climbing the tree and killing it by destroying the upper foliage—the crown.

When early European explorers praised ponderosa pines, calling them regal, swooning over the sun-dappled parklike groves that were so easy to travel through on horseback and sometimes even by wagon, they were celebrating a forest thoroughly shaped by wildfire. Not only fire in its more extreme form as scorching walls of flame marching across the mountains,

---

PREVIOUS PAGES **An air tanker drops fire retardant to slow the advance of the Chelan Complex Fire, which consumed nearly 89,000 acres just south of Chelan, Washington, in September 2015.**

Landscapes in the West have evolved in ways that deeply reflect the powerful force of wildfire.

but also the far more frequent low-to-moderate-intensity regular burns foresters today call stand-maintenance fires.

When stand-maintenance fires occurred, burning through many western woodlands roughly every ten to fifteen years, what got burned up were the lower branches the trees had dropped and the occasional toppled tree, along with needles, cones, and small plants of the forest floor. Meanwhile, if diseases or insects came along and wiped out patches of mature trees, as those trees fell they left holes in the canopy. In little time, young pines would begin sprouting in those holes and vying for dominance, eventually replacing the fallen trees and closing the canopy again. On it went like this, to a greater or lesser degree, often for centuries at a time.

But beginning in the latter half of the 1800s, use of the western forests by humans greatly intensified. Big changes followed. Grazing livestock, for example, altered the plant communities on the forest floor, as the hooves of cattle provided seedbeds in which clusters of young, shade-tolerant trees could gain purchase. This alone helped to make the forest a more crowded place—something that would have major consequences when fires came along. In addition, massive logging operations got under way, with loggers felling mostly the largest, oldest trees; what remained were tighter bunches of young-to-middle-aged timber. But probably nothing had more of an effect on the fires we see burning today than the fact that in the early twentieth century we began eighty years of fervent effort to put out every fire we could get our axes and shovels and saws on.

More about that later. But for now you may be wondering how we know what the forests looked like before Europeans showed up and how we've gained knowledge of the way wildfires have shaped what we see today. The answers are held in some painstaking scientific research, research that involves going to the trees themselves, rediscovering what many indigenous cultures around the world have pointed out—namely, that trees are the keepers of stories about the landscape.

A first step in unraveling the burns of long ago is to examine a given section of forest for so-called fire scars. Whenever sufficient heat from a passing fire penetrates the bark of a tree, it causes death in a section of the underlying cambium layer; from then on, those dead cells show up as a lesion, or fire scar. Hot fires on the ground likewise cause damage to certain sections along the base of a tree—scars that will last either until the tree rots or some other damage occurs. By studying a number of trees in a small area, scientists can piece together information about the size of past fires, the direction they were traveling, even the fuel load on the forest floor at the time of the burns.

Once this fire scar data is collected, it's compared to annual tree ring records, which provide a yearly account of the general growing conditions across the lifetime of an individual tree. In this way fire scars can be dated—not just to the year they occurred but often to the very month. Thus the fire history of a forest, even an entire mountain range, can be reconstructed with surprising accuracy. These kinds of historical records, in turn, have allowed us to get a better sense of the degree to which both the frequency and intensity of fires have increased over the past couple of hundred years.

FIRE SEASON IN THE AMERICAN WEST HAS SURELY CHANGED. WHEN it comes to heat, drought, and the resulting inevitable flush of wildfire, the new millennium has brought a run of trouble-filled summers. Trouble at the edges of inland cities in central and southern California, where flames have rolled in from loose clusters of live oak and chaparral to devour hundreds of homes. Trouble too in the sweet-smelling Engelmann spruce and Douglas-fir groves outside ski towns like Vail and Jackson and Big Sky. Trouble on cattle ranches from Washington to Wyoming, where fires have roared out of the foothills to consume pasture and fences and barns. And trouble especially in a thousand forested subdivisions stretching from the

Indigenous cultures have called trees the keepers of landscape history. These fire scars in a ponderosa pine trace wildfires across some three hundred years.

western slope of the Sierra Nevada to the eastern slope of the Rockies, from the Sangre de Cristo Range of New Mexico to the Chugach Mountains of Alaska.

Even for those to whom wildfire has become familiar, it's hard to comprehend the collective impact these burns are having on the landscapes and communities of the West. Not that long ago, 4,000-acre fires were a big deal. But now firefighters are routinely called out on burns three or four times that size. "I think about the big fire season in California back in 1987," says highly experienced firefighter Matt Holmstrom, superintendent of the Lewis and Clark Hotshots in Great Falls, Montana. "And then the next year, massive burns in Yellowstone. They seemed like such exceptional years. And now they're pretty much routine."

To even begin to gain a realistic view of what fire season in the western United States has become, it might help to put our imaginations to the task of conjuring up what all this might look like in any given year—in fact, what it *has* looked like in most years since 2000—from a vantage point high in the sky. Think for a moment of climbing aboard a small plane in late July or August and taking off from southern California, maybe from the Orange County airport, lifting off the runway to begin a low, slow flight across the region.

Shortly after leaving the ground we might spot wildfires to the east, in the San Bernardino and San Jacinto mountains, and near Ventura, too, maybe started in dry grass by an off-road vehicle, or by campers, or even by a faulty power line—the bigger ones already churning enormous billows of smoke that will be seen some five hundred miles to the east, in Arizona. To the north, meanwhile, other burns might be easily visible in the Sequoia National Forest, these perhaps ignited by dry thunderstorms and then fanned into infernos by day after day of hot winds. And on it might go, mile after mile as we fly north, past Chico and Red Bluff and Redding, beyond the flanks of Mount Shasta, where flames could well be crowning in the big pines of the Shasta-Trinity and Six Rivers national forests.

All along our journey we'll be able to find clues about the conditions that have allowed such fires to burn in the first place. We'll see telltale "bathtub rings" around the upper reaches of reservoirs, each one—in the wake of so little rain and less-than-average snowfall—holding only a small fraction of its normal storage capacity. There'll be big whirls of dust, too, kicking up from grasslands near Yuba City, from baseball fields in Stockton and Chico, from the dozens of dirt roads that lie twisted like cooked spaghetti across the state's sprawling complex of national forests.

TOP The Sayre Fire in November 2008 consumed 489 homes in the Sylmar section of Los Angeles.

BOTTOM Clouds of smoke loomed over Colorado Springs on the first day of the June 2013 Black Forest Fire, which killed two people, destroyed more than five hundred homes, and forced the evacuation of thirty-eight thousand people.

And if our plane dips low enough, wherever there are homes near these forests and shrub lands, wherever rural neighborhoods and subdivisions and ranchettes are found—and those number in the tens of thousands across the West—we'll surely see people scurrying about moving sprinklers around to wet the perimeters of their yards, sprinklers that in some places have been going nonstop for days. Others, meanwhile, might be lifting their noses, checking for the smell of smoke; looking skyward every hour or so, scanning for any promise of clouds that might deliver rain. Stopping too every so often to glance up at flags atop courthouses and schools and post offices—knowing well that when it comes to wildfire, their fortunes are tethered to the wind.

As we travel on into the Oregon Cascades, the mountains near Santiam Pass might well be so engulfed in smoke that from our small plane it's like looking down through massive clouds of fast-moving, yellowish-gray fog. Helicopters with water buckets might be buzzing along the edges of the burns, avoiding the main plumes because of the severe turbulence, dropping loads and then heading off to refill at streams located well away from the densest smoke.

Much the same in Washington, north and south of Snoqualmie Pass, as well as to the east, in North Cascades National Park near Lake Chelan. Several dozen more burns might be whipping along the spine of British Columbia, too; and many more still in Alaska, where not dozens, but hundreds of forest landscapes might be burning simultaneously. While most of the blazes deep in the Alaska wilderness burn unchecked, others—located near cities like Anchorage or in places like Kodiak Island—are fought aggressively. At several points we're likely to catch sight of stocky DC-10 airplanes roaring along just 300 or 400 feet above the ground, dropping at each pass more than 10,000 gallons of scarlet-colored retardant ahead of the flames.

Heading southeast, dropping back into the continental United States at the Idaho border, we'll find the region smoke-laden, the fires ranging from 50 acres in the Idaho Panhandle to more than 3,000 acres in the nearby

A helicopter drops retardant on the Harris Fire, one of more than nine thousand fires that burned in California during 2007. By the time that fire season had finally come to an end, more than a million acres had gone up in smoke.

Kootenai National Forest in Montana. Traveling down through western Wyoming into Colorado, we might catch flames leaping across the Grand Mesa Plateau, and well to the south in the Weminuche Wilderness near Durango; all of that, in turn, might be mixed in with big smoke drifting in from northern Arizona and southern Utah. From here, heading east, we'll hopscotch across still more wildfire, maybe in the West Elk and Hunter-Fryingpan wildernesses, on Fossil Ridge and the Collegiate Peaks.

If we finally land the plane at some small airstrip in southern Colorado, catch a ride to town, and start walking the streets, it won't take long to notice that in a lot of places conversations too are being lit by fire. Nervous talk about nearby burns. But also about the lack of rain in the forecast, about how it's not cooling off at night like it used to. About the growing list of closures on the nearby national forest. In truth, here, just as in so many other

Increasingly, the much-loved panoramas of the American West are infused with smoke. Yet there's much more than scenery at stake here. In a single week during September 2015, the federal government spent more than $240 million battling out-of-control wildfires.

parts of the West, drought and fire and heat have been driving people's conversations for months. In March they focus on the low snowpacks; in early May, on how the pasqueflowers and lupine are blooming weeks earlier than they used to; and in June, with every day that comes and goes without rain, on talk of weather. Not that these are by nature anxious people. But wildfire, and the dry conditions that have been hanging over the area for years, is weighing heavy on their minds.

Poignant scenes in such communities from California to Colorado have taken place in July and August of the drought years, in people's driveways, as hundreds of men and women and kids have carried boxes and suitcases from their houses to the families' cars or pickup trucks, making ready to drive away with precious pieces of their lives should word come down that wildfire is closing in and the time has come to evacuate.

And so it goes across the West. Life in a land of flames.

THE CLOSER WE LOOK, THE CLEARER IT'S BECOMING THAT WILDfire—which has long exerted an enormous impact on western lands—is becoming a bigger force than ever before. Between 2000 and 2015 an astonishing ten fire seasons saw more than a dozen so-called megafires—a term still somewhat loosely defined but these days often applied to burns of more than 100,000 acres. There have been four years in the last half century when more than 9 million acres have burned in the United States, and all of them have been since 2006. Little wonder that in more than half the years since 2000 the Forest Service has drained its entire wildfire suppression budget, shoveling funds into what is often now an annual expense of $2 billion. There's a line firefighters often toss out: that 99 percent of the burned land comes from 1 percent of the fires. In North America, megafires actually constitute fewer than 3 percent of the fires but currently account for more than 90 percent of the area burned. Make no mistake about it—big fires are becoming common fare.

And fire season is growing longer. Research spearheaded in 2015 by South Dakota State University found that worldwide, from 1979 to 2013 warmer temperatures and more rain-free days caused the length of the wildfire season to increase by 18.7 percent. (Notably, while the amount of precipitation falling around the world is essentially the same, these days that moisture is coming in fewer days. Which of course leaves more dry days when conditions are right for wildfire to occur.) As for the western United States, a review of Forest Service records shows that the wildfire season today is longer than it was in 1972 by an astonishing seventy-five days—about ten weeks. There are a couple of reasons for this. First, all things being equal, western forests become prone to fire ignition within a month of the snowpack disappearing, and today that snowpack is melting several weeks earlier than it did in the 1970s. At the same time, hot and dry conditions are persisting longer in the fall, so the fire season is also longer on the back end.

A big question many Americans are asking these days isn't so much

The Rim Fire in and around Yosemite National Park consumed more than 253,000 acres between August and October 2013 and is the biggest fire on record in the Sierra Nevada. Such megafires are becoming more common.

why wildfires are burning per se, but why they're burning so hot and fast. If the size and scope of those fires were easier to grasp by means of our imaginary trip in that small plane, the factors that make them so frequent and severe are probably best unraveled with feet on the ground. Were we to walk through the ponderosa and Douglas-fir forests near the place we landed in southern Colorado, for example, on the west side of the Front Range, on first glance they might look much as they did in the 1980s. Besides dark

Heavy smoke pours from a wildfire that ignited in late October 2003 in the Simi Hills of southern California, then burned into November and consumed 108,204 acres and thirty-seven homes.

sprawls of conifers are clusters of aspen and wild rose, fireweed and lupine, currants and grouse whortleberry, part of the weave of life growing here for centuries. Chickadees are still scouring the aspen branches looking for bugs, while red-tailed hawks circle in a blue, blue sky above, undaunted by the gusty winds, scanning the ground for mice and voles. If anything, the land today would first and foremost look weary, tired well before the time of year when things should be tired—dusty and withered, many of the leaves of the deciduous trees chewed and frayed by bugs.

But if you'd spent recent decades on foot or horseback in this country, like many of the locals, you'd have more than vague perceptions about the

place being merely parched. Just up the trail another mile or so the branches of thousands of conifers have gone to cinnamon brown—signs of a pine bark beetle infestation coursing through the forest like an ocean wave, an event that in this area alone has already claimed thousands of acres of forest. Even many of the still-green trees in these woods are actually sick, their trunks lined with telltale pitch tubes, created as the trees have tried to eject the beetles inside. Knowing the fate that awaits them, some people have taken to calling them ghost trees.

With a warming, drying climate, not only are insect populations surviving better throughout much of the West, finding homes in drought-stressed

conifers, but also with warmer winters there's been a doubling of the rate at which they multiply. So in this drainage, and in a thousand others, in the near future more trees are going to die: killed while standing, then five to seven years later falling to the ground during windstorms, like a great tumble of jackstraws.

As a rapidly changing climate rocks well-established ecosystems, new conditions are altering forests and shrub lands in unprecedented ways. While weather and climate are of course incredibly dynamic, Earth's ecosystems have in general gone through long periods of relative stability. That, in turn, has allowed much of the life in those ecosystems to live successfully in defined niches. And while it's true that many species have proven capable of changing as their niches change (at least outside of rare catastrophic events like volcanic eruptions), those adjustments typically occur over long periods of time, often thousands of years. An alarming thing about climate change is how quickly things are shifting, moving at a pace much faster than some species can adjust to.

For example, up in the alpine zones, the mountainous regions above timberline, the little pika—the smallest member of the rabbit family—is already disappearing from much of its former range, crippled by warming temperatures and resulting vegetational changes at high elevations. At the same time, as the subalpine forest climbs higher and begins occupying what is now the alpine zone, the white-tailed ptarmigan—a bird that depends on tundra, defined as treeless plains at high altitude—could also disappear, at least regionally. Indeed, by all indications a considerable number of species will be either highly stressed or gone altogether within the next fifty to a hundred years.

Climate change, combined with heavy fuel loads from decades of aggressive fire suppression, has led to bigger, hotter wildfires. The Wallow Fire in 2011 was the largest in Arizona history, burning more than 538,000 acres in Arizona and New Mexico.

Changing too in the Rockies are the numbers of amphibians historically found here—creatures such as leopard frogs and tiger salamanders—as the wet places in the hollows of the hills are increasingly lost to rising temperatures and diminishing precipitation. Meanwhile, in some creeks cutthroat trout are faltering, no longer able to make their annual spawning run due to drying streambeds.

As the ground has gone drier, native foxtail and bluebunch wheatgrass are disappearing, yielding to more drought-tolerant invasive species like cheatgrass and loosestrife, both of which have big advantages over native plants. Following a wildfire, cheatgrass has the ability to produce twice the amount of root material that native bluebunch wheatgrass can manage. And as usual, pulling one thread in nature's web of life tends to disturb other threads. Because these exotic, invading grasses are less nutritious, grazing animals like elk may end up giving birth to lower-weight calves in the spring; that, in turn, could translate into lower survival rates. And fewer elk calves could in time mean fewer predators, including coyotes and bears and mountain lions, as well as fewer scavengers like eagles and foxes. At the same time a proliferation of nonnative grasses in the Southwest, including buffelgrass in the Sonoran Desert, has created conditions that lead to hotter, longer-lasting fires. More important, because many of the native plants of this region didn't evolve with such intense fire events, in the end they may not be able to survive and will ultimately disappear.

Yet for all this ongoing scrambling for survival among myriad species, what millions of people all over the western United States tend to think of first when someone mentions climate change, is wildfire. Natural as fire may have been to the creation of these landscapes, it's become a much bigger, far more frequently occurring force. What's more, this is happening at just the time when more and more human communities are moving into those regions of the West most vulnerable to wildfire.

The Tea Fire began near Santa Barbara, California, on November 13, 2008, and was fanned by 85-mile-an-hour winds blowing down the Santa Ynez Mountains. It would take five days to contain and would destroy more than two hundred homes..

SINCE THE 1980S, LOW-DENSITY RURAL HOUSING DEVELOPMENTS have mushroomed throughout the West. Such housing has long been known in wildfire circles by the rather lackluster term of wildland-urban interface, or WUI. In simple terms, WUI is a name given to any fire-prone region in the United States where wildland fuels intermix with human structures. These are everywhere: near the edges of state lands and national forests, along national park and monument borders, on BLM lands, next to national seashores and historic parks, and also on private ranches and woodlands and conservation preserves. Of the 2.3 billion acres that make up the United States, about 1 billion, or 44 percent, of those acres are considered wildland-urban interface. Of that, about 220 million acres—an area roughly twice the size of California—is now designated by state and federal managers as being at high risk of wildfire.

But there's a lot more to the story than that. Almost 40 percent of new development in the western United States is taking place in the WUI. Since

1990 the rate of conversion of wildlands to wildland-urban interface has grown on average at the astonishing rate of 3 acres a minute, 4,000 acres a day, 1.5 million acres a year. Eight million new homes have gone up in the WUI, until by 2015 these lands contained about forty-six million single-family homes and hundreds of thousands of businesses. Roughly 120 million people, or more than a third of all the people living in the United States, are in the wildland-urban interface. And perhaps most incredible of all, more than 80 percent of these lands are still to be developed.

While wildland firefighters in the United States are spectacularly successful at what they do, keeping 97 percent of the roughly one hundred thousand fires that break out every year to fewer than 10 acres in size, the ones that get away, that get big and destructive, are usually in the WUI. In the 1960s wildfire was destroying on average about two hundred structures every year, but these days burns are taking between three and five thousand. In 2015, forty-five hundred homes burned in wildfires. Since the year 2000, almost forty thousand homes and businesses in the WUI have been lost. A million more—property worth about $140 billion—are now considered to be at high or very high risk.

But there's something else to note here, too. While we often think of western communities and subdivisions as being at risk from wildfire (in some areas home values have dropped near wildfire sites), less often do we consider the fact that such housing makes it a lot more likely that fires will happen in the first place; indeed, one early study in California suggested

TOP **With every passing decade, the risk of catastrophic losses from wildfire in the wildland-urban interface continues to grow. The 2015 Valley Fire, in Lake County, California, devastated Middletown and surrounding areas, destroying nearly two thousand structures. There were four fatalities.**

BOTTOM **The spring 2015 Tomahawk Fire, in the northeast section of Camp Pendleton, California, caused large evacuations both on and off the base.**

Fires burning in the Flatiron Mountains outside Boulder, Colorado, threaten homes in the wildland-urban interface and impair the air quality in Boulder.

that in the populated portions of the state, only 4.2 percent of destructive wildfires were caused by lightning. Plain and simple, the more houses, the more likely a wildfire will happen.

And it's not just property that's increasingly at risk as wildfires rage hotter, longer, and closer to human habitat. It's air and water quality as well. The clean, fresh air long touted as a primary attraction of the West—for residents and tourists alike—can in some months be hard to find. And the smoke from wildfires can affect air quality far from the burn itself; for example, smoke from fires in Arizona and New Mexico in June 2011 affected air quality for large areas east of the Rockies and extended up into the Great Lakes region.

While for generally healthy people wildfire smoke is mostly just an unpleasant nuisance, for those with health problems such as asthma or heart conditions the burns can be a serious problem. Smoke from forest fires is made up of a fairly complex mix of gases, including carbon monoxide, as well as fine particles of various sizes. It's those particles that cause most of the trouble—leading to burning and itchy eyes, to headaches and runny noses, and when they enter the lungs they can cause serious problems like bronchitis. For those living with cardiac disease such as angina and congestive heart failure, or with such pulmonary diseases as asthma and emphysema, there are often more serious consequences still.

Children too are more at risk in the face of wildfire smoke—both because their respiratory systems are still developing and because pound per pound they breathe significantly more air than adults do. Even the unborn can be affected. Studies in southern California following major fires in Los Angeles and Orange counties in 2003 found decreased average birth weight among infants exposed to wildfire smoke in utero, with the greatest effects coming to those whose mothers were in heavy wildfire smoke during the second trimester. While these birth-weight reductions don't appear to be an immediate health threat, that could change as wildfires become more common.

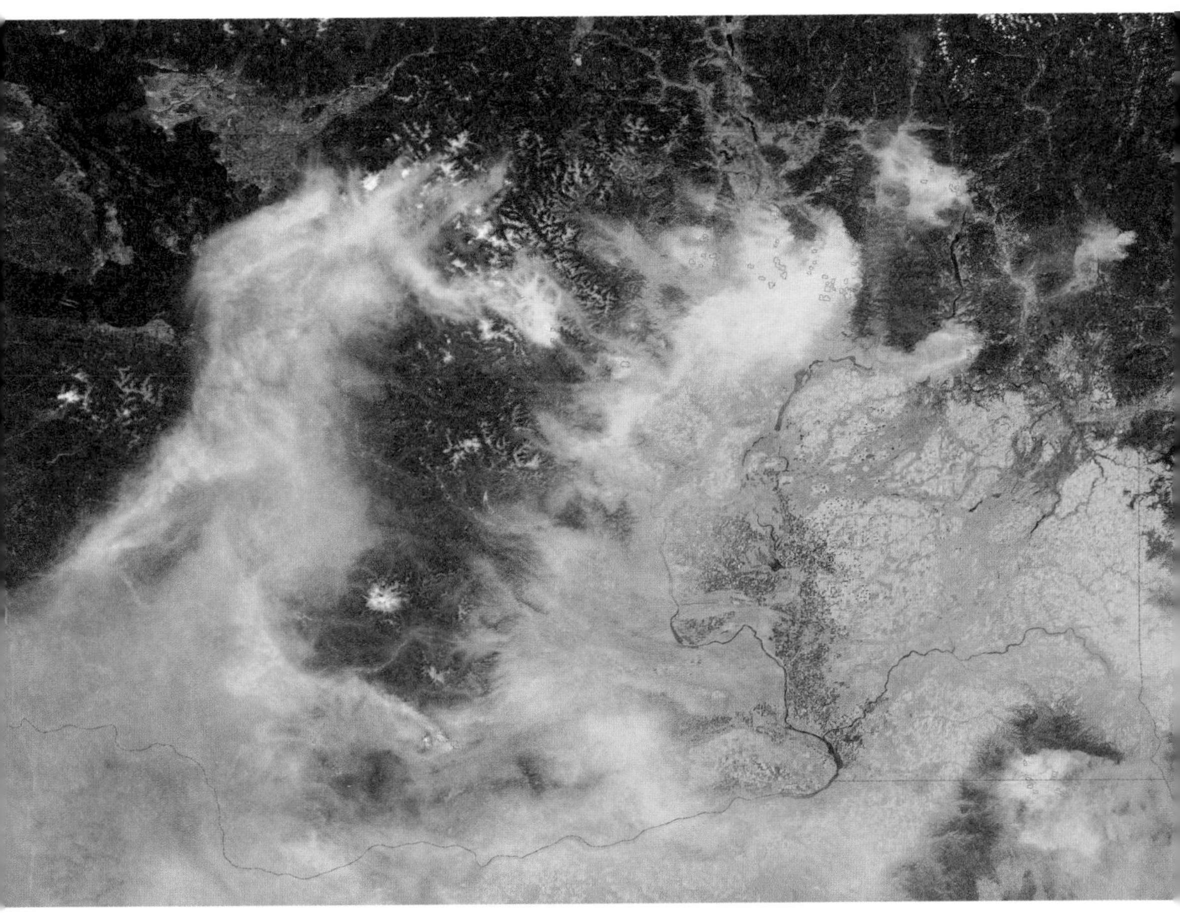

An August 2015 NASA satellite image of wildfires in Washington State shows heavy smoke covering most of the communities around Puget Sound.

Increasingly, residents of the western states living near active wildfires are being told to be aware of air quality index numbers, which are routinely reported to local weather feeds from regional air quality monitoring stations. These are in turn shared in newspapers and on radio, television, and the Internet. In fact, it's possible to roughly gauge health threats with your own eyes.

In otherwise clear, noncloudy conditions, if you can see 10 to 15 miles, the air quality is generally good.

A view of less than 10 miles means the air quality is beginning to deteriorate.

Seeing less than 5 miles on an otherwise clear day means the air is unhealthy for young children, adults over sixty-five, pregnant women, and people with asthma or heart or lung disease. Outdoor activity should be limited.

Visibility of 3 miles or less equates to conditions that are generally bad for everyone. The groups of people listed above should avoid all outdoor activity.

Finally, if visibility is a mile or less, the air may be hazardous. Every person, no matter how healthy, should avoid all outdoor activities. Windows and doors should be kept closed. Air conditioning units need clean filters and should be run only with the fresh air intake controls closed. People should avoid using anything that burns, such as gas stoves and candles. (Even vacuuming is a bad idea.) And as for the paper dust masks at the hardware store, those are meant to trap large particles like sawdust; they're of no use when it comes to keeping fine smoke particles from entering the lungs.

Beyond smoke, big fires—especially in mature forests, where fires have traditionally been suppressed—tend to release temporary pulses of heavy metals into the air and water. Along the Front Range of the Rockies in Colorado, for example, following the Hayman Fire in 2002 some places saw lead, mercury, arsenic, aluminum, and cadmium levels increase by factors of twenty-five hundred. Nitrate levels in streams are also directly linked to the severity of burns, with big fires being capable of producing concentrations forty times higher than normal. Dissolved oxygen levels in waterways also increase under such circumstances—not only acidifying streams but also decreasing the effectiveness of community treatment facilities.

LET'S RETURN TO THE PONDEROSA AND DOUGLAS-FIR FOREST IN southern Colorado where we imagined a wildfire igniting at the start of this chapter. In the times we live in now, that story will unfold in completely different circumstances from a fire occurring in the same place many hundreds of years in the past. Today, a stand-maintenance fire of the kind that originally shaped the forests might, because of fire suppression, have been absent for many decades; in addition, years of drought in our changing climate would've led to tinder-dry conditions on the ground. And finally, in the mouths of the drainages that wind down and out of the high, rugged backcountry where the fire starts, as well as along many miles of foothill country that lie between those valley stems, there are likely clusters of homes built by retirees and refugees from urban living, people who love the out-of-doors and breathtaking mountain scenery.

That's a microcosm of conditions all over the West that are giving fire season a completely different meaning and character. How disruptive all this proves to be will depend in large part both on the speed at which the climate changes and how intelligent—and nimble—we humans are when it comes to reacting to these new realities. We can perhaps best prepare ourselves by understanding more about conditions on the ground, and about how fires start and behave, topics we take up in the chapters to come.

KINDLING

The era of suppression meets the age of drought

IN THE WARM MONTHS OF 1910, PORTIONS OF THE AMERICAN WEST experienced what was up until then one of the most severe fire seasons of historical times. The first burn that year kicked off in April, in Montana's Blackfoot National Forest, followed by hundreds of fires across the region—some started by loggers, others by campers, and more than a hundred more by steam locomotives. The Forest Service was young and hugely understaffed, created just five years earlier by President Theodore Roosevelt, and the season would test the grit and pluck of its rangers unlike any other. By August so many fires had broken out in Montana, Idaho, Oregon, and Washington that the fledgling agency found itself not only signing up every able-bodied citizen it could find to wield an ax or shovel but also urgently petitioning then-president William H. Taft for Army troops—a request Taft refused before he finally relented and sent in some four thousand soldiers.

As a result of heroic effort, by the third week of August it seemed the most threatening blazes were well in hand—so much so that on August 19

Beginning in the early twentieth century, heavy fuel began building in the forests as we suppressed nearly every wildfire.

forest supervisors began releasing firefighters. But on August 20, under the powerful slap of hurricane-force winds, the embers from earlier fires suddenly blew back to life in what would later be called the Blowup, creating walls of flame hundreds of feet high, gobbling a phenomenal 3 million acres of trees, forcing full-scale evacuations of towns across the northern Rockies. The upward draft of the smoke columns was so fierce that flaming pine trees were literally plucked from the ground by the roots and sent spinning across the landscape. A forester named Edward Stahl called the fire "a veritable red demon from hell."

By the time the fire began to settle—just three days later, on August 23—under a curtain of light rain and snow, it had killed eighty-six people,

This stand of white pine in Idaho's Coeur d'Alene National Forest was flattened by the Big Burn of 1910.

burned down much of the Idaho town of Wallace, and destroyed enough timber to fill a freight train more than 2,400 miles long. It was by some measures the worst wildfire in American history. Forever after it would be known simply as the Big Burn.

This was the fire that would bring to the public eye a forty-two-year-old former ranch foreman and miner-turned-forest-ranger named Edward Pulaski. While Pulaski was supervising firefighting crews south of the town

of Wallace, near Placer Creek, he and his men were suddenly overwhelmed by massive walls of flames. With death hot on their heels, at the last minute Ed managed to herd his crew of forty-five into the abandoned War Eagle Mine. Inside the tunnel was heavy, choking smoke and little oxygen, forcing the men to lie prone on the ground to gulp what air was available, choking and coughing as their throats were burned. Yet as hellish as conditions in that tunnel were, Pulaski knew the mortal danger waiting outside the tunnel entrance; at one point he drew his pistol and threatened to shoot anyone who left. Miraculously, all but five of the firefighters survived the ordeal.

What makes Pulaski near and dear to the heart of almost every firefighter today, however, is his invention the following year of a simple firefighting tool now known as a Pulaski. Consisting of a hardened steel head that has an adze on one side and an ax on the other, it's a near-perfect implement for digging, scraping, and chopping on the fire line. Within a decade after its invention, the Forest Service in the Rockies had fully adopted the tool, contracting for it to be manufactured commercially. By the 1930s it was in use across the nation.

But back to the Big Burn. At a time when the very existence of the Forest Service was being seriously questioned by many in Congress, that conflagration gave the agency an opportunity to argue its value in the fight against wildfires. Miners and loggers didn't care for taking on that particular job, after all, and neither did the railroads. And while even back then some were pushing to let fires in the wilderness burn themselves out (and even to do prescribed burns, taking a cue from longstanding practices of Native Americans), others, including future Forest Service chief Ferdinand Silcox, argued that with enough manpower and scientific prowess, the Forest Service could eliminate fire from the landscape. And in the end, his was the view that prevailed.

The argument may even have saved the Forest Service, which for the next sixty years touted itself in large part for the role it was playing in

In 2011 at the Albuquerque Balloon Fiesta, Smokey Bear was still going strong, making his wildfire prevention effort the longest-running public service campaign in American history.

protecting the nation's timber through wildfire suppression. To get a sense of just how strong this commitment to putting out wildfires was, consider the existence of the so-called 9:00 a.m. rule—a directive that said that if a wildfire broke out, anyplace and at any time, the goal of the Forest Service (though of course one not always achieved) was to have the blaze under control by nine o'clock the next morning.

Thus began our many-decades-long era of wildfire suppression. Wildfire, said land management agencies, was simply bad—a message they felt was critical to get out to the public at every chance. To that end, the Forest Service set about communicating this idea about the menace of wildfire through educational campaigns.

A 1944 poster introduces the original Smokey Bear, doing what he does best.

ENTER SMOKEY. SMOKEY BEAR (TECHNICALLY NOT SMOKEY *THE* BEAR, by the way, as some call him) first appeared complete with his trademark ranger hat and denim pants in a 1944 poster; he looks especially strong and fit, pouring a pail of water on a burning campfire, standing above the words "Care *will* prevent 9 out of 10 forest fires!" Smokey came into being as the replacement for a deer named Bambi, who Walt Disney had loaned to the Forest Service in the fall of 1942 for a year-long fire prevention campaign.

Created as a character by art critic Harold Rosenberg, then brought to the canvas by Albert Staehle, Smokey Bear gained even more status in 1950, when a real black bear cub was found in a forest fire on the Lincoln National Forest in New Mexico, his coat singed, his paws and hind legs burned.

Brought back to fire camp and nursed to health, the little bear was first called Hotfoot Teddy, then later Smokey. Once the papers got wind of the story, the little bear was an instant hero. In time he landed at the National Zoo in Washington DC, where he lived for twenty-six years. Confusion about his name—Smokey Bear versus Smokey *the* Bear—likely started in 1952, when songwriters Steve Nelson and Jack Rollins wrote what became a very popular tune about the famous bruin, with the chorus "Smokey the Bear, Smokey the Bear / Prowlin' and a'growlin' and a'sniffin' the air."

It would be hard to imagine a more effective sales bear. Even today Smokey is recognized as a firefighting champion by more than three out of four kids and a staggering 95 percent of adults. He has managed to stay on point for decades, dutifully delivering the message of the Forest Service and countless other land management agencies, showing up in massive television campaigns and Little Golden Books, always painting wildfire as a foe to be vanquished.

WHICH OF COURSE IT SOMETIMES IS. BEING A CREATURE OF THE woods, though, Smokey might well have known what it would take humans longer to understand—that fire can in a lot of cases be a blessing. It can create conditions for understory plants like fireweed, asters, and clover, and does much to support elk populations, which frequent recently burned areas to feast on protein-rich grasses. (In Yellowstone in the years following the 1988 fires, elk not only dined on the grasses but now and then even consumed the burned bark of lodgepole pines, something that doesn't generally happen with unburned trees.) And fire can clean away the downfall that piles up over time without regular clearing burns—downfall that after enough years have passed can pile up to turn what would've been average fires into major conflagrations. Ironically, the most significant result of suppressing all wildfires has been to create extraordinarily flammable forests.

Plain and simple, because of our earlier fire suppression policies, a staggering 300 million acres of western forests are today suffering from unnaturally heavy fuel loads.

Suppressing all fires has had other unfortunate consequences, too. For example, tree-boring insect populations, as well as various tree-killing infestations like mistletoe and blister rust, have gone unchecked. Thus when fires do get loose in the landscape, they're fed by massive quantities of insect- and disease-killed timber. Often entire forests burn, including mature trees, which typically have an ability to stay alive through all but the most severe wildfires. This also has had the regrettable effect of creating monoculture forests—woods made up of nearly all the same tree species, with trees essentially all the same age. Besides such forests having less diverse plant and wildlife populations, they are loved by insects like pine bark beetles, which can reproduce year after year in these large homogenous tracts, moving from one wedge of timber to the next.

In some places aggressive fire suppression has changed the makeup of the forest itself. Across vast reaches of California, for example, overeager firefighting has led to forests being populated with a lot more shade-tolerant white fir than they otherwise would contain. Ironically, one reason people were fighting forest fires at every turn through much of the twentieth century was to protect commercial timber. But as it happens, white fir is of little interest to the timber industry due to its relatively weak, knot-filled wood. What's more, unlike ponderosa pine, white fir retains its lower branches as it matures. These branches serve as ladders for wildfire, so that fire kills not just the white fir itself, but in California, sometimes also neighboring giant sequoias.

Wildfire has other beneficial effects that fire suppression can short-circuit. Most of the world's terrestrial plants gain the minerals they need to grow—including calcium, nitrogen, potassium, and phosphorus—by absorbing them through their roots. Those minerals aren't available by accident; they end up in the soil thanks to a wide range of essential organisms called decomposers, a category of life that includes bacteria, fungi, and earthworms. But in much of the West—especially the intermountain West—cool, dry air slows down decomposition to a tiny fraction of what it is in warmer, wetter places like the South or the coastal Northwest. What's more, pine needles, which cover the floor of a great many interior western forests, have waxy coatings that make them resistant to decay, which slows decomposition even more. When trees fall over in these sorts of climates, they can lie there for decades. So in these regions, a major way nutrients get put back into the soil is through wildfire.

What's more, if a forest is deprived of stand-maintenance fires, if all the branches and fallen trees and ground vegetation that would've normally burned don't, when fires do occur they burn faster and hotter—sometimes, in fact, hot enough to compromise the soil nutrient cycle. While a small ground fire can reach temperatures of 1400 degrees F, sporting flames 3 to 5 feet high, big fires in places with heavy fuel loads can be a sizzling 2000 degrees F, with flames 150 feet high. And again, because these more severe burns often reach the crowns of the trees, something that in the past didn't happen all that often, entire woodlands can be consumed.

One of ecology's fundamental principles is that plant energy, or biomass, is not static. Eventually, in every system it must either decompose or combust. Clearly, decisions about fire suppression were made before most people had wrapped their heads around this fundamental principle, because to think we could put out every fire was to believe we could keep biomass building up forever. It never dawned on us that one day we'd have to pay the piper.

And even those who long ago understood there would be days of reckoning, pushing for prescribed burns even in the early twentieth century, never imagined the reckoning would be so enormously amplified by climate change. They didn't know the fuel load would be increased dramatically

In 2013, the lightning-started Alder Fire burned more than 4,000 acres on a peninsula at the south end of Yellowstone Lake in Yellowstone National Park. The warming, drying climate, along with heavy fuel loads due to fire suppression, are the most important factors affecting wildfire risk.

due to drought-stressed trees—at least in part because so many trees have been killed in recent years by pine bark beetle and other infestations. They couldn't have anticipated the fact that today there are considerably more frost-free days in the West than there used to be; indeed, the growing season in the Northwest and Southwest is now two to three weeks longer than it was in 1960. That, in turn, has led to a much greater profusion of grasses coming up in the spring—grasses that dry out by midsummer, becoming nearly perfect tinder for wildfire.

WE KNOW FROM TREE RECORDS—ESPECIALLY TREE RINGS, WHICH show general growing conditions—how much more severe than previous centuries our own times are in terms of drought. Indeed, if we were to pick a weather theme for the American West in the new millennium, we'd be hard pressed to find one more fitting than this.

First there were the massive water scarcities of 2000 to 2004 throughout western North America, bringing the driest conditions in some eight hundred years, or in other words, since roughly the middle of the twelfth century. For example, according to the Bureau of Reclamation, inflow into Lake Powell from the upper Colorado River Basin during that period was about half of what's considered average; and the 2002 inflow was the lowest recorded since Lake Powell began filling in 1963. Crops withered in the field. Trees died by the tens of thousands, including ponderosa in Arizona; pinion and juniper in New Mexico, Colorado, and Utah; and spruce and Douglas-fir throughout the West.

Then the years from 2008 to 2011 brought astonishingly dry conditions to California, Texas, and the Southwest—in several places rivaling the crushing conditions of the Dust Bowl. The years from 2013 through 2015 brought even worse circumstances, especially in California, when that state set all-time low precipitation records and Sierra snowpacks were abysmal.

In the midst of all this twenty-first-century drought, scientists have begun puzzling out what proportion of these extraordinary conditions can be chalked up to climate change, and some preliminary answers to that question are emerging. Researchers at NASA, the Lamont-Doherty Earth Observatory, and the University of Idaho have pored through wind speed,

Devastating drought gripped the Northwest in 2015. Water in Detroit Lake, in the Oregon Cascades, reached its lowest level since the reservoir was constructed sixty years earlier.

temperature, and soil moisture records, tree ring samples, and precipitation data. Using as their baseline the climate variations not just of the last century but of the past thousand years and aided by some extremely sophisticated computer models, they've been able to suss out the contribution made to drought by human-caused climate change (so-called anthropogenic warming) versus natural climate variation.

According to these early investigations, anthropogenic warming may bear as much as 27 percent of the responsibility for the megadroughts now plaguing the West. More provocative still, by all indications such severe droughts are now more than twice as likely to occur as they were a hundred years ago. And looking to the near future, things aren't likely to get better. According to these and other data, in the second half of the century, residents of California and much of the Southwest will have an 80-percent chance of being in a severe drought that will grind on for thirty to thirty-five years.

WHILE A NUMBER OF INSECT SPECIES—AND PATHOGENS TOO, LIKE blister rust—are able to take advantage of heat-and-drought-stressed forests, in the West none has had a more significant effect than the pine bark beetle. It's not that beetles haven't long been a natural part of western forests. In the past, though, their numbers were held in check in part by the fact that winter temperatures in these regions fell to minus 20 or 30 degrees F, which limited the ability of the insects to spread. But these days such temperatures are in many years nowhere to be found. In the relative comfort of a warming climate, beetle populations have exploded, moving ever farther north, reaching parts of Canada where they've never been recorded before. British Columbia has so far lost some 65,000 square miles of timber—an area roughly the size of Wisconsin. Warmer temperatures have also lessened the amount of time it takes at high elevation for the beetles to go from egg stage to maturity, slashing that development process from two years to just one. And while healthy trees can often produce enough resinous pitch to flush out beetles as they try to enter the tree, drought-stressed conifers are less able to do so. In the Rocky Mountain states, beetle infestations claimed

---

TOP With sustained drought in California, finding water resources for fighting fires has been an increasingly difficult challenge. Reservoirs, for example, have been at such low levels that helicopter pilots sometimes struggle to fill their drop buckets.

BOTTOM Pine bark beetles have killed massive numbers of trees in British Columbia. In recent years in the American West, more trees have been killed by beetles than by wildfire.

FIRE
WATER
FIRE
WATER

more trees in the period from 2000 through 2015—more than 45 million acres—than they did in the entire twentieth century.

To understand the reason so many trees are being killed by beetles and disease means stepping back a bit to consider the direct impact of drought on trees. Scientists refer to the climate-related dryness that leads to tree weakness and death as water balance deficit. This deficit is defined as the difference between potential evapotranspiration, which is the amount of moisture capable of being drawn out of trees by things like solar radiation, temperature, and wind, and actual evapotranspiration, which refers to how much moisture is available. When the potential evapotranspiration exceeds the amount of water available, there is said to be a water balance deficit. As researchers struggle to get better and better at predicting wildfire activity, an increasing number are leaning on models that focus on this water balance deficit. Unlike large-scale climate models having to do with processes like El Niño, the attraction of water balance models is that they're far more useful for understanding what's going on in a specific forest.

Consider that during the first of several major droughts that have unfolded in the new millennium, between 40 and 80 percent of the pinyon trees died across some 4,600 square miles in Arizona, Utah, Colorado, and New Mexico. While it's true that the drought caused extreme stress to those trees, leaving them vulnerable to beetle infestations as well as to attacks by an insect called the five-spined ips, some uncertainty exists about what actually caused the trees to die. Some researchers figure it had to do with carbon starvation, where a stressed tree stops photosynthesizing, starving itself of the carbon dioxide needed for that process, as the stomatal openings of the needles close off to prevent water loss.

Others, meanwhile, blame a pumping failure. As you might imagine, getting water up a tree, especially a fairly tall one, is no easy task. Trees take advantage of ground moisture by pushing water molecules upward by means of a slight pressure exerted by the root system; those molecules then

As conifers are increasingly killed by beetle infestations, lost too are the foods they provide to birds and animals, whether Clark's nutcrackers or grizzly bears.

rise through a capillary-like network known as the xylem. But that's only half the story. At the same time, as water passes through leaf membranes into the air (a process known as transpiration), it causes a slight upward pull on fluids in the xylem. In a drought, when there's little water in the ground to be had, the pressure needed by the tree's capillary system to keep water flowing up and out to the branches increases considerably. Sometimes it gets high enough to cause something known as cavitation—essentially a break in the xylem tube from the cells filling with air to maintain osmotic pressure and then bursting. While a tree can handle some cavitations, with too many it can no longer lift the water it needs to survive.

For a long time scientists were mystified about why some trees in the Southwest succumbed to these mechanical failures during that period of severe drought, while other trees—including large swaths of pinyon and juniper in the Gila Wilderness—did not. The answer may have something to do with differences in soils. The soils of the Gila Wilderness and other parts of southwest New Mexico tend to be deeper, more nutrient rich, and also of finer texture than those in the northern part of the state. That finer texture allows them to store water as ice and then release it as conditions warm; the coarser, sandier soil up north, meanwhile, tends to lose water either by evaporation or by letting it sink to depths where it's unavailable to the trees.

The much-talked-about 1950s drought in this region, which also led to large die-offs of pinyon trees, mostly affected older trees on dry, lower-elevation sites. The more severe recent drought episodes, by comparison, have had impacts almost everywhere—low elevations to high, and across all age classes. Even junipers, which are among the most drought-tolerant species of trees, have in recent years been dying by the thousands. Some researchers think the problem may have been made worse by the fact that beginning in the mid-1970s and continuing for a couple decades, precipitation was significantly above average. That wet period may have created forests of big trees that required more water, resulting in increased competition when the droughts of the new millennium came along.

THE DEGREE TO WHICH DROUGHT AFFECTS A FOREST IS DETERmined in large part by where the forest is located. For example, in the Pacific Northwest, a major predictor of big fires is a drought starting early in the winter and persisting all the way through spring and into summer. Absent that kind of protracted dry period, it's pretty hard to get the landscape west of the Cascades to burn. In part this is because forests closer to the coast have an abundance of woody plants, both living and dead, and those plants

The beautiful pinyon pine is now at great risk from drought throughout much of the Southwest.

retain moisture for the entire forest ecosystem, which in turn makes it difficult for fire to spread. In addition, a closed forest canopy, where the treetops essentially touch each other, will shelter whatever fuel is on the ground from the drying effects of sun and wind.

But if you cross the Cascades into the dryland forests of eastern Oregon and Washington, or into the Rockies, or even south into much of

Dried grasses added to the fuel load that was ripe for burning in the Lincoln National Forest in New Mexico when the Little Bear Fire broke out in June 2012. The fire was contained at 4 acres with a fire line around it, but then wind blew embers from a torching tree outside the fire line. The fire eventually burned 242 houses and 44,330 acres.

California, it takes a lot less time for the land to dry out when precipitation flags. Indeed, there may be abundant moisture all through the spring into early summer, but if there's a slight snowpack and a couple of months of no rain, the area will be at very high risk of burning. Also, in many places, such as Colorado's low-elevation ponderosa forests, fires are more likely following wet springs, in part because early grass growth leads later to dried grass that is easily ignited. In grass-dominated ecosystems, such as those of the Great Plains, fire incidence is actually much higher in wet years than in dry.

Forests in places with less regular rain experience a much slower rate of decomposition; basically that means more dead wood on the ground, which can be especially troublesome when you couple it with the aggressive fire suppression employed for so many decades. And while dead fuels can slow a fire when they're wet, they of course have no ability to either draw water from the ground, as a live plant can do, or to close off cells to reduce moisture loss in times of drought.

LAND MANAGERS THROUGHOUT THE WEST HAVE BEEN TRACKING drought and fuel loads in forests—and the resulting risk of wildfire—for a very long time. Every spring through fall, these managers collect fuel samples, small pieces of wood from both live and dead trees in public and private forests. These samples are then placed in airtight containers and transported to regional labs, where technicians calculate fuel moisture. Fuel moisture is determined by comparing the weight of the wood at the time it was collected to its dry weight, which is measured after drying the samples in special ovens.

Measurements are done on a variety of vegetation sizes, from less than a quarter inch thick (like dried grass) up to 6-inch-diameter trees, and each size is classified according to the number of hours it takes to respond to changes in atmospheric moisture. For example, a fuel such as a log from a dead lodgepole pine that when totally saturated by rain requires roughly a thousand hours to dry out again and be ready to burn is classified as a thousand-hour fuel. Dried grass, on the other hand, needs just an hour and so is routinely referred to as a one-hour fuel. Fuel moisture levels are assigned for each of these categories. As an example, after two months of severe drought, the moisture level of the thousand-hour dead fuels in a ponderosa pine forest might be just above 10 percent—drier than kiln-dried lumber.

High fuel moisture numbers mean that if a fire starts, its advance will be slowed by the fact that much of the energy of the flames has to go into drying out the water in the wood before it can ignite; low moisture levels, meanwhile, mean a fire will spread rapidly, especially if pushed by wind.

These measurements, then, form the bedrock of how fire danger is assessed across the nation.

THIS SAMPLING OF VEGETATION IS STILL THE MAINSTAY OF HOW land managers determine fuel loads, allowing them to determine fire danger in specific locations. This information, sometimes combined with thermal images from NASA satellites, allows the staging of limited firefighting resources in a way that makes possible the most efficient deployment possible.

At one time, agencies concerned with fire suppression in the United States had no fewer than eight different approaches to calculating fire risk; given that each agency was using different measurements, talking and acting from the same page was difficult at best. Attempts to address the problem started back in 1959, when researchers began developing a complex set of equations—combining measurements of fuel moisture, fuel types, weather, and other factors—to calculate fire danger. These danger stages, in turn, were then related to staffing levels, thereby linking fire risk to specific, detailed management decisions.

Today, in any given week during fire season, state and federal land managers report potential fire danger in their regions by use of one those previously mentioned equations, now part of the National Fire Danger Rating System. Some, for instance, will use something called the energy release component (ERC), which measures the available energy, or potential "heat release" per square foot at the flaming head of a fire front. As live trees lose moisture, as in times of drought, and dead wood dries, the ERC gets higher.

As various inputs are entered into the rating system, the potential risk of fire breaking out and the potential ways in which it might behave if it does are expressed using several different measurement categories. The "spread component," for example, combines data about fuel moistures, wind speed, and slope of the ground to predict how many feet per minute a fire is likely to spread. The "ignition component," on the other hand, places a numerical value on the likelihood of an ember or firebrand causing a fire that would require suppression. The "burning index," meanwhile, estimates how hard it might be to control a burn based on how hot it is, as well as the speed at which it might be moving.

Put all these together and you can get some remarkably helpful guidance—direction that allows for the staging of important resources based on fire risk in different parts of the country as well as knowing where to put the arrow on those Smokey Bear signs outside of ranger stations, on the dials that range from "low" to "extreme." The complexity and dynamism of this rating system is extraordinary, with hundreds of people constantly working to refine it for their particular landscapes.

For example, after the 1988 fires in Yellowstone, researchers figured out that the "flammability threshold" for lodgepole pine forests in that particular national park occurs when fuel moisture in thousand-hour fuels drops to 13 percent; at that level, fires start easily and spread quickly. Meanwhile, the park's Douglas-fir forests, located mostly in the northern part of Yellowstone, typically rise not from floors deep with pine needles, like lodgepoles, but thick mats of green vegetation; in addition, Douglas-fir forests typically have far fewer downed logs on the ground, which means less fuel to spread the fire. As a result, that particular ecosystem has earned an entirely different potential fire rating on the National Fire Danger Rating System scale. As conditions escalate across the fire season, managers are increasingly able to make sound predictions about which forests are likely to burn first.

GIVEN ALL WE KNOW NOW ABOUT THE IMPACT OF DECADES OF FIRE suppression, does Smokey Bear still have a role? In fact, he does, especially when you combine increased fire danger from frequent and prolonged drought in the West with the rapidly growing number of homes in the

The Forest Service today recognizes that controlled burns in fire-dependent ecosystems can make the forest healthier and reduce the risk of catastrophic wildfire damage.

wildland-urban interface. Smokey is still busy educating the public about wildfire, in part through a website and a Facebook page. You can see a map of current wildfire incidents on his website, as well as read about how the Forest Service perspective on fire suppression began to shift after the Yellowstone fires of 1988 and the fire season of 2000. Today the agency gives itself the option of deciding "to allow lightning-caused fires to continue to burn in areas that will not affect the safety of people while reducing fuels."

It also recognizes that by reintroducing fire into fire-dependent ecosystems in a controlled setting, it's possible to recreate the positive effects of natural fire while at the same time preventing catastrophic losses from uncontrolled wildfire.

It's a bit ironic that as we've at last come to a deep understanding of the role of fire in maintaining healthy forests, climate change is so altering conditions on the land—creating thousands of acres of insect kill, as well as otherwise stressing and killing trees—that in many places, especially anywhere even remotely close to the wildland-urban interface, the inherent wisdom of controlled-burn policies is having to be reconsidered because of the heightened danger of major conflagrations.

Which in a sense makes Smokey's job more important than ever before.

COMBUSTION

From spark to flame to firestorm

MILLIONS OF PEOPLE ALL OVER THE WEST HAVE SINCE THE TURN of the twenty-first century taken to holding their breath when summer thunderstorms roll in during drought conditions, fearing lightning strikes will set the land ablaze. Yet in truth, nearly nine out of ten wildland fires are human caused. Some come from cigarettes tossed out the window, or from campfires, or from spreading structure fires. Other times fires that were lit to burn brush piles or other debris get out of control; they can even start days after such controlled burns are over, igniting from hot embers still in the soil (these are called holdover fires). Then there are fireworks, as well as loose sparks tossed off by machinery and downed power transmission lines. And finally, sometimes wildland fires are the result of arson. In fact, of the more than 120,000 fires that break out in the United States in an average year, more than 10,000 are thought to be set intentionally.

The science of figuring out how a wildfire started is known as pyro-forensics. Pyro-forensic investigators are driven by the desire not just to assign

Lightning strikes during a summer thunderstorm in Arches National Park, Utah.

responsibility for human-caused burns but also to guide education and fire management efforts when it comes to preventing accidental human-caused ignitions. Like any detective work, the job requires a significant amount of sleuthing—beginning with a thorough combing of a burn area from the outside edges in, usually along carefully laid grid lines.

A pyro-forensics expert uses clues called fire indicators, or directional indicators, to point ever backward to the original point of ignition. In a nutshell, figuring out the direction of the burn, combined with the topography of the land and wind direction, ultimately leads an investigator to the place where the fire began. For example, in a given area usually one part of the fuel load, such as standing trees, remains unburned, which orients the investigator to the direction the fire was moving. Likewise, sometimes extreme heat can "freeze" leaves on the branches of trees, leaving them pointing in the direction of the winds created by the fire. Also, more soot gets deposited on the sides of branches and vegetation facing the oncoming fire, while green leaves tend to curl in the direction the heat is coming from. Fires tend to burn outward in an expanding V, where the point indicates the place where ignition occurred. Once that ignition area is identified, and we're often talking about a piece of land smaller than a hundred square feet, it's time to get down on hands and knees and sift through the ashes, looking for anything from a cigarette butt to a lighter.

Even if such a culprit is found, though, there's more work to be done. Did the fire start by accident, or was it set intentionally? Could power lines have been down in the area? Could roadwork or firewood cutting have been happening, allowing a machine to toss off a spark? To answer those questions, investigators must go through the often tedious process of tracking down and interviewing neighbors, hikers, campers, even first responders—all in a careful search for witnesses. They may have to check satellite images, too, or even contact the Federal Aviation Administration to establish whether pilots of any low-flying planes that were in the area could provide helpful information.

The newest addition to wildfire crime scene investigation, used so far in only a handful of states, is the so-called arson dog. First investigators isolate scent from a fire's origin point—either through physical evidence or by placing dirt or other ground material onto sterile gauze pads. This is then presented to the dog, who is able to detect and subsequently track skin cells shed by a human body. During the busy wildfire season of 2012, the Bureau of Land Management and the Forest Service in Idaho used a bloodhound named Jesup to investigate both the Avelene and Whitehorse wildfires.

Increasingly, states are handing out harsh consequences to those who start fires, even by accident. People have been fined tens of thousands of dollars and sometimes jailed for starting fires with chain saws or lawn mowers or by burning brush in unsafe conditions. Two men whose abandoned campfire caused the 2011 Wallow Fire in Arizona, which burned 841 square miles, were ordered to pay $3.7 million in restitution. A Forest Service worker who set the Hayman Fire in Colorado in 2002 served five years in federal prison and owes more than $59 million in restitution. Power companies have also paid millions in fines. Beyond what their insurance covered, San Diego Gas and Electric coughed up about $420 million for the role of their system of power poles and transmission lines in starting the Witch Creek, Guejito, and Rice Canyon fires in 2007—blazes that killed two people and destroyed more than thirteen hundred homes. Likewise a jury in New Mexico found two utility companies mostly at fault in the 2011 Las Conchas Fire—the largest blaze in that state's history until that time (a record broken the following year).

IT'S TRUE THAT MOST WILDFIRES TODAY ARE STARTED BY HUMANS. And for thousands of years, indigenous cultures intentionally set fires—both to stimulate the growth of certain desirable plants and to improve grazing conditions for wildlife. Yet a much older and in general far more unpredictable source of ignition in western forests is lightning. How do lightning strikes occur?

Let's return to the mountains of southern Colorado in August, the setting for the lightning strike that opened this book. As is true on most summer days, over the course of the morning the cool ground gets warmed by the sun. That warm earth, in turn, heats the adjacent air; and warm air, being lighter or less dense than the cooler air above, rises into the sky through late morning and during most of the afternoon. As these thermal rafts of air get higher, they cool. And because cool air can't hold as much moisture as warm air, whatever vapor those rafts of air hold condenses and

A Forest Service engine crew battles the 156,000-acre Las Conchas Fire on the Santa Fe National Forest in New Mexico. The fire started on June 26, 2011, when a gust of wind blew an aspen tree onto a power line.

OPPOSITE The Springs Fire, which burned more than 6,100 acres of the Boise National Forest, was one of twenty-six human-caused fires in Idaho during the summer of 2012.

forms drops. Those drops, being heavy, fall back toward the ground, on the way passing through other cells of warm air still on the way up.

But from that simple convection process comes something rather difficult to pin down—something a bit mysterious even today. The collisions those falling droplets are having with the rising warm air causes them to gain electrical charge—sometimes a negative charge, other times positive. Positively charged droplets, as it happens, tend to end up at the top of the clouds, while negative ones gather at the bottom. Now if this splitting of

positive and negative happens to a sufficient degree, in time it can lead to a charge being induced either cloud to cloud or downward to the earth below. And quite often that strong urge to equalize the electrical charges leads to what we know as lightning.

On average, lightning hits planet Earth hundreds of thousands of times a day. The spears themselves boast temperatures of about 50,000 degrees F—some five times hotter than the sun. Given that dry wood can burst into flames at under 600 degrees F, it seems pretty remarkable that out of hundreds of lightning strikes that hit a given forest in any year, only a small percentage lead to wildfire. In fact by some estimates, of the thousands of ground strikes that occur annually in the West, only between 1 and 4 percent lead to ignitions. The reason for this is that while the intense heat of a lightning bolt may damage individual trees, most of the time lightning simply doesn't generate current for long enough periods to start fires.

But those statistics, like so many others, might be changing in the face of climate change. Not only is the land becoming a lot drier, making what combustion from lightning strikes does occur more likely to spread into wildfire, but lighting itself is getting more common. Some researchers predict a 12-percent uptick in the number of strikes for every 1.8 degrees F (1 degree C) rise in temperature. And though it's not at all clear just where on Earth these extra strikes will happen, at the rate things are changing, by the end of the century there could be 50 percent more strikes than there are today.

COMBUSTION SCIENTISTS (NOT TO MENTION PLENTY OF JUNIOR high school science teachers) talk often of the fire triangle, or the three ingredients needed in order for something to burn: heat, fuel, and oxygen. When wood initially comes into contact with a heat source—something

This tree in the Sierra Nevada was struck, as so often happens, without being burned.

in the neighborhood of 300 degrees F—the first thing that happens is that the cellulose, which is the main component of a plant's cell walls, begins to break down. As that happens the wood gives off gases, including hydrogen, carbon, and oxygen, which become visible as smoke. At around 500 degrees F, the molecules of those compounds break apart, at which point atoms recombine with oxygen to form, among other things, water and carbon dioxide.

And that's about when ignition occurs—when fire happens—through a process called pyrolysis. Pyrolysis is a kind of chain reaction in which the heat given off by the burning gases causes more gases to be released, and they too then catch fire. Again, for the burn to keep going it needs sufficient heat, fuel, and oxygen. Likewise—and this is the big key to fighting wildfires—if you deny the burn any one of those things, it will go out. What's left over after the fire, that thing we call ash, is made up of those minerals in a tree or plant that simply don't burn, including things like calcium, magnesium, nitrogen, and potassium.

How much energy is needed for ignition depends in part on the size of the object being heated. The reason we use small pieces of wood, or kindling, to start a campfire or light a fireplace is that it takes a lot less energy to raise a sliver of wood to the temperature needed for pyrolysis than it does to do the same thing with a 6-inch-thick log. Also, the wetter the wood—from rain or snow or simple humidity—the bigger the heat source has to be, because the water must be driven off before ignition can occur.

NOW LET'S IMAGINE THAT NOT QUITE SEVENTY-TWO HOURS AFTER that bolt of lightning first spears the ridge in southern Colorado and causes dried grasses and beetle-killed branches to ignite, flames have marched up the mountain to consume 600 acres of ponderosa pine and Douglas-fir. The fire is moving upslope at high speed, fanned by wind, advancing nearly half

Combustion happens in the presence of heat, fuel, and oxygen—the fire triangle.

a mile with every passing hour. The biggest worry is that it will spot into a canyon to the north, where, given the harsh winds, it could make a run toward a group of homes. If our goal is to contain wildfire when it threatens life or property, we need to understand how wildfire moves across a landscape and grows. With that kind of information in their pockets, today's wildfire incident commanders can better anticipate fire behavior, making assessments of which homes or other structures are going to be most at risk of burning and applying available resources to best match current conditions.

And this brings us to another, somewhat less familiar, triangle known by wildland fire scientists and firefighting crews. This one has to do with the three main elements that determine how a wildfire will behave. The first factor, fuel, we're already familiar with; simply put, there needs to be something to burn. The second is weather—be it the blustery afternoon wind that drives the burn or the fact that a stretch of rainless weeks has left the wood of the forest dry as lumber. The final element is topography, or terrain, which has a great deal to do with determining both where a fire goes and how fast it gets there.

Setting those basics aside for a moment, it's important to note that there's really quite a lot we don't know about the workings of wildfire. This is partly because funding for wildfire science has historically been limited, what with the vast majority of dollars going to suppression. And at the same time, wildfire is incredibly complicated—especially when compared to other combustion systems like jet engines or heating devices. Wildfires, after all, have no defined boundaries. They change by the second, driven by an incredibly complex mix of variables in fuel and weather and topography. Long story short, those who study wildfire combustion have to deal with lots of moving parts.

Also, the men and women who study wildfire have made their share of wrong turns. Research chemical engineer Bret Butler, at the Fire Sciences

Laboratory in Missoula, Montana, points to landmark studies conducted from the 1960s through the 1980s that measured the heat inside of fires with simple wired devices called thermocouples. "We thought any thermocouple would give accurate temperature measurements. But as it turns out, that's not really true." For one thing, he says, such readings were highly dependent on the size of the thermocouple being used; as the size of the device increased, it responded more slowly to changes in temperature. Which is sort of a big deal, because the flames of a fire have enormous fluctuations. In a word, they pulse. "For us to think there was a single temperature in a flame was a huge error," says Butler. "We were getting an average. But an average of what?" Butler also points out that past wildfire research focused much of its effort on radiant heating (energy emitted by electromagnetic waves from a single source, like the sun), while today there's a recognition that convection is a big player. Convection has to do with the fact that molecules of gas near the fire gain heat energy, then travel through the environment to another location (say, an adjacent blade of grass or pine needle), giving off their energy and, under the right circumstances, causing ignition.

So let's drill down a little more into the factors that allow wildfire to spread. Watching a raging canopy fire sweeping through a forest, it's easy to assume that burns travel by virtue of the tips of the flames jumping from treetop to treetop—one great tongue of flame lighting a set of branches and then still another in a kind of spectacular leapfrogging of radiant heat. But some rather remarkable research conducted in 2015 at the Missoula Fire Sciences Laboratory has turned that notion on its head.

As part of the USDA Forest Service's Rocky Mountain Research Station, the Missoula Fire Sciences Laboratory employs scientists and technicians in cutting-edge wildland fire research. It's the only facility in the world dedicated exclusively to studying wildfire through controlled experiments in burn chambers—investigating fire ecology, the processes of fire, and such questions as how smoke disperses across the landscape and how best to manage fuel loads. Much of their work has already resulted in safer, more effective wildland fire management.

For a great many years, scientists there and at a few universities across the country have been trying to understand the spread of fire by lighting flames in those special burn chambers—devices resembling small wind tunnels filled with different kinds of flammable natural materials, from grass to small logs. But because there are so many variables in those fuels (and one thing that makes scientists squirm is too many variables), occasional spot flaring in the burn chamber had to be written off as the result of extra resin in the wood, or random clumping, or even slightly different degrees of dryness. Anxious to tame that sort of unpredictability, researchers at the lab instead started burning precisely shaped pieces of cardboard. And when they did, more or less by accident they unearthed a never-before-understood, yet absolutely vital, principle of fire behavior.

It's helpful to think of a flame front as an undulating line of peaks and troughs, looking something like the crude blade of a saw. What scientists in Missoula were able to finally see was that between every set of peaks in the flames, there occurred in the intervening trough a pair of rapidly spiraling vortexes carrying extremely hot gases down and out. Perhaps it was these downward-spiraling gases, they reasoned, that in many cases caused fires to advance—causing ignition not by radiation, where material near the peaks of the flames ignites, but through convection, happening down low, at the level of the grass, pine needles, and other fuels on the forest floor.

Excited by what they discovered in the Missoula lab, the scientists teamed up with researchers at the universities of Kentucky and Maryland to conduct real-world experiments outdoors. In the process they burned all sorts of things, from trees to fields of grass. And sure enough, there it was again, the same behavior they'd witnessed in the lab—swirls of smoke and flame driving to the ground to become the leading edge of the advancing fire.

Combining this basic understanding of how fire often advances with

This vignette from the 2012 Whitewater-Baldy Complex burn in New Mexico, the largest wildfire in New Mexico history, seems to bear out the Missoula lab finding that fire advances by way of downward-spiraling gases at the level of fuels on the forest floor, not by jumping from treetop to treetop.

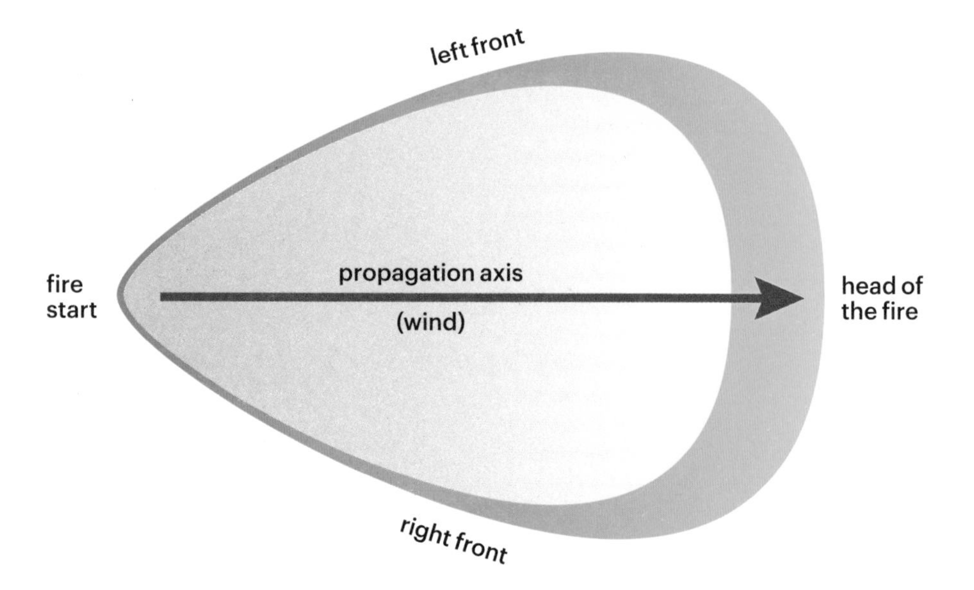

This model of wildfire progression shows how a fire typically burns most aggressively at the head of the fire (often driven by wind) but also expands along the sides, creating the shape of a pear.

information about the types of fuels present will in years to come help us assess the potential for wildfire to spread, as well as calculate how fast the spread may happen. At the same time, understanding this business of "ground advance" may also help us better protect rural housing developments by applying more effective treatments to the surrounding forest floor.

AS A WILDFIRE ADVANCES, AS IT GETS BIGGER, OTHER FACTORS from that second fire triangle routinely used by firefighters become increasingly important—namely, terrain and weather. For starters, sloped ground, which is a very common feature of the American West, allows fire to travel by "climbing." Flames, which of course burn in an upward direction, are closer to vegetation growing on the upslope, which means they dry, preheat, and ultimately ignite those fuel loads more easily than the ones lower down. If the fire is burning in steep mountainous terrain and the amount of air supplied to the burn is restricted by a steep slope above the flames, this can lead to what essentially is a small low-pressure zone. Sometimes without any warning the fire will leap up into those low-pressure zones, rapidly expanding in both size and strength.

Fire also climbs because upslope and up-canyon winds—common in mountains in the afternoon, as the land warms—tend to push fires upward. Conversely, downslope winds come at night, as the land cools. Along with lower temperatures and higher humidity, downslope winds can help firefighters by pushing the fire back on itself, onto ground that's already burned.

By studying regional records from the past we've been able to figure out that certain local landscape features, like ridgetops or cliff walls, may in moderate fire conditions be a reliable buffer against the spread of fire across the landscape. But under extreme drought conditions, those buffering features of the landscape have done little good at all. And those features become more important the bigger a burn gets, in large part due to their ability to shape wind speed and direction.

Of all the factors that affect fire spread, none is a more constant worry to firefighters and homeowners alike than wind. Besides bringing extra oxygen to the blaze, which increases its intensity, strong winds can send sparks and embers and flaming pinecones out well ahead of the main burn, causing a rapid advance by means of "spotting." Gusty winds are especially troublesome, since those kinds of winds not only come on fast but also can change direction suddenly, putting firefighters at big risk. In the right terrain and with heavy dry fuels, a blustery wind can cause an almost unimaginable rate of spread.

This was the case in the Lake County, California, Valley Fire of 2015. The fire started early in the afternoon of September 12, grew to 10,000 acres by that evening, and incredibly, reached 50,000 acres by the following day.

Lots of sloped terrain allowed fire to climb in the fast-burning Rim Fire of 2013 in California's Stanislaus National Forest. It took fire crews nine weeks to contain the blaze, though remarkably, it would be another year before the fire was completely out. It burned more than 400 square miles and was the third largest wildfire in California history.

More remarkable still, in April 2016 a cluster of wildfires in the tar sands region of northern Alberta, Canada, pushed by fierce winds, grew by more than 100,000 acres in twenty-four hours, burning more than a thousand buildings and leading to the evacuation of some eighty thousand people.

A study done by the Los Angeles County Fire Department in the early 1970s found that the burns in the previous decade that had grown into the largest wildfires—back then defined as at least 5,000 acres—all happened on days when the average wind speed was more than 30 miles per hour. It's quite possible that climate change will in some places result in even stronger winds in the future, making wildfires more dangerous still. Unfortunately, at this point science is a long way from understanding, let alone predicting, how and where such wind increases will occur.

Still another factor that has to be considered is humidity. Hot air absorbs more moisture than cool air, so higher air temperatures translate into lower humidity. Also, because wood either loses or gains moisture depending on what's happening in the surrounding air, low moisture means greater flammability in the forest. This is why, all things being equal, a wildfire raging during the day often slows as night comes on, since night brings greater humidity and lower temperatures (along with a diminishment of upslope winds), offering firefighting crews a fresh chance to make progress containing it.

FIREFIGHTERS WELL KNOW THAT ERRATIC, UNPREDICTABLE THINGS can happen close to a major burn besides those related to weather and terrain. These have to do with dynamics created by the fires themselves. Really big wildfires require enormous quantities of air. Roughly speaking, for every unit of biomass being burned, feeding the combustion takes about 6,000 units of air. So strong is this need for air supply that researchers have measured the winds at the base of a wildfire—so-called in-drafts—at a phenomenal 60 miles per hour. And those in-drafts can cause a fire to move in very unexpected ways.

Yet another atmospheric danger with wildfire—one responsible for a considerable number of firefighter deaths—has to do with the smoke cloud or column that forms above large fires. (When you see such a column above a wildfire, incidentally, light-colored smoke generally comes from grasses and shrubs, darker smoke from heavier fuels, and very dark gray or black smoke is usually a sign of burning structures.) Columns of rising heat in the sky above a large fire tend to produce pyrocumulus clouds—lacking in rain but capable of producing extremely strong, gusty winds. A mature smoke

Firefighters can tell much about a fire long before they get on the scene, just by looking at the smoke. Dark smoke indicates heavy fuels, while lighter-colored smoke indicates fuels made up of grasses and other small plant materials.

cloud may have vortex winds so strong it can rip soil, hot embers, fire retardant, even entire burning trees right off the ground and carry them into the air. When the central column of this pyrocumulus cloud finally collapses, it can have deadly results for those working on the ground below.

Finally, when the air above a burn is unstable, something called a fire whirl can develop—a phenomenon where flames begin to draw upward in a spinning motion, forming a strong vortex, sometimes several hundred feet across, not unlike a tornado. When firefighters see fire whirls they know to use extreme caution, as these are telltale signs that conditions are right for the fire to act suddenly in completely unexpected ways. In addition, fire whirls can rise and travel considerable distances, in the process dropping

A pyrocumulus cloud develops above the Oregon Gulch Fire, started by lightning 15 miles east of Ashland, Oregon, at the end of July 2014.

burning embers onto the flammable landscape beyond a fire crew's location and pinning them between the primary and secondary burns. Spotting from fire whirls is also extremely dangerous to communities, as the falling embers they carry have the potential to ignite homes and other structures.

THIS BRINGS US TO THE QUESTION OF HOW OUR KNOWLEDGE OF the biophysics of wildfire behavior can be applied to predicting what will happen in any particular wildfire incident. Experienced firefighters today still use a simple yet highly useful calculation tool developed way back in the late 1800s known as a nomograph. In a nutshell, a nomograph is a graph-based, simple pen-on-paper method of making complex calculations, which

This pyrocumulus cloud viewed across Yellowstone Lake was produced by grasses, shrubs, and trees burning in the 1988 fire.

in the case of wildfire includes enabling assessments of likely fire behavior. By plotting fuel moisture and wind speed on the graph, for example, it's possible for a trained fire behavior analyst to determine rate of spread and flame height—critical information when it comes to figuring out what resources will be required to fight the burn.

But thanks to the rapid advance of computer modeling, firefighters also have available to them an expanding bevy of sophisticated models for predicting wildfire growth and behavior—models powerful enough to handle

A fire whirl developed on Rabbit Mountain, Oregon, during the 48,000-acre Douglas Complex Fire, which was started by lightning in July 2013.

thousands of inputs in real time, creating relatively quick predictions of what's likely to happen in the coming hours and days of a given fire event. Some of these models are for use in the field, on a laptop or even a cell phone, to help incident commanders draw on historic fire data to better manage the burns in front of them. There are even apps being tested that allow, in the midst of a burn, instant assessment of which structures are most at risk of catching fire.

It was only in the 1960s that researchers started building mathematical models for predicting wildfire. The earliest efforts were launched by the staff of the Missoula Fire Sciences Laboratory, who came up with formulas for linking the direction and speed of wildfire advance to wind, vegetation, and terrain. Slowly but surely, that early work has led to more high-level models. In past years the most commonly used tactical modeling system has been FARSITE, which anticipates fire movement by combining weather data with data on wind, fuel, and terrain, displaying forecasts on a simple two-dimensional map.

FARSITE is still in nearly constant use by firefighters from Maine to Alaska to California. And all in all, it works pretty well. But given the increasing complexity of fire—more extreme atmospheric conditions due to climate change, and more intense localized fire weather due to bigger, hotter burns—and also given other information that incident commanders increasingly want to be able to put their hands on, FARSITE can feel at times like using a slide rule to plan a trip to the moon.

A significant limitation of the FARSITE program is that it assumes a degree of uniformity that doesn't exist in the real world, from fuel being evenly distributed across a given forest to static weather conditions—plus it lacks the ability to anticipate weather generated by the fire itself. The program also can't account for the recent breakthroughs in understanding how fire travels nor accurately anticipate spotting and dangerous events like fire whirls. These limitations create lots of room for surprises; and in the world of firefighting, surprises can be dangerous, even catastrophic. (Like in 1994, when a predicted cold front arrived with enormous bluster at a burn outside Glenwood Springs, Colorado, surprising fourteen firefighters who died in the blaze.) In addition, because programs like FARSITE are anchored in historic weather data, they haven't proved very good at accommodating climate change factors, from drought levels to wind events—which means they often underestimate how far the fire will spread in a given number of days.

FARSITE and nomographs are now being supplemented by other, more sophisticated prediction tools. For instance, at the National Center for Atmospheric Research in Boulder, Colorado, scientists are working to more fully link the complex, often self-reinforcing loop between how a fire

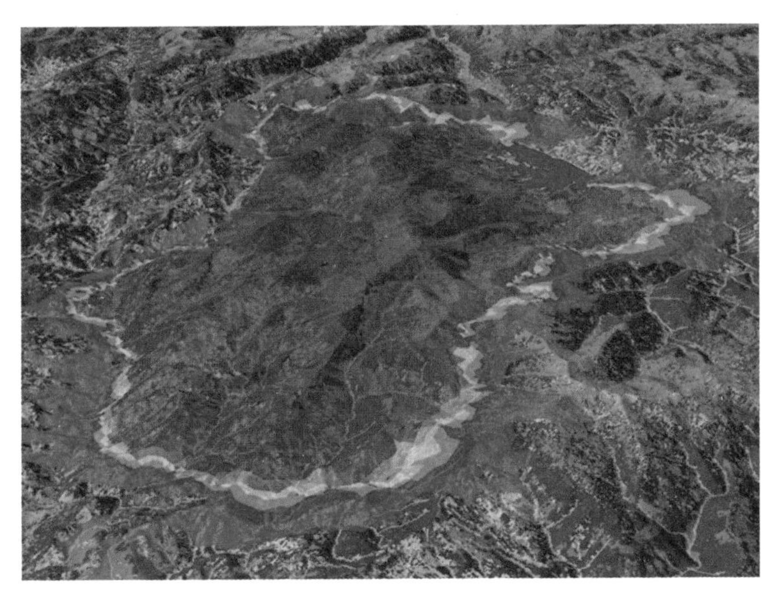

A FARSITE image simulates the likelihood of fire spread in a section of Idaho's Boise National Forest. The fire has a lower probability of spreading, indicated by the colors blue and purple, to lower elevations. Such probability information can be of great help to fire managers, allowing them to better understand risk to life and property.

behaves and changes in the atmosphere. The rising air at the leading edge of a fire can pull flames up ridges and canyons, and sometimes even across entire canyon slots. Likewise, established local wind patterns can have big effects on a burn, especially in times of severe drought, creating the kind of rapid forward or sideways movement that can frustrate suppression efforts. The more such localized terrain and weather trends can be factored into wildfire models, the more effective and safer will be the tactical decisions made on the ground.

Yet as helpful as these more sophisticated models can be, they cost millions of dollars to develop. And at least for now, most research money is being spent instead on figuring out the effectiveness of different landscape treatments, such as prescribed burns and thinning operations, as well as how to better stabilize soils and vegetative communities in the aftermath of burns to avoid erosion and landslides.

THERE'S ANOTHER PROBLEM WITH THESE SOPHISTICATED MODELS. That's the fact that most firefighters simply don't have enough computing resources in the field—nor will they probably have them for decades—to run complex predictive software. And that points us to yet another cutting-edge project at the Missoula Fire Sciences Laboratory, a modeling tool with the rather fanciful name of WindNinja. The fruit of fifteen years of research, this model is designed to be used on a laptop computer by firefighting teams in the field and may prove a breakthrough in estimating how fast a fire will spread over a given time period.

Until now, wildland firefighters have had limited tools for figuring out the role played by wind, during the fire and also afterward when trying to analyze what happened. They could take direct wind speed and direction measurements by walking the area with a Kestrel, a handheld metering device. "But then you've got the problem," explains Bret Butler, "of trying to extrapolate that point measurement into a bigger scale." They could also turn to weather service forecasts. But such forecasts offer just one wind direction and wind speed number for every 4 to 10 kilometers (1.5 to 3.8 square miles). And while that's helpful, it can often be impossible to apply a scale that coarse to specific drainages.

"What we've done," says Butler, "is to develop a model that predicts wind speed and direction essentially every 100 to 200 meters [109 to 218 yards] across the landscape. We approached the problem as mechanical engineers—figuring that surface wind flow in complex mountainous terrain would be a lot like airflow over an aircraft wing or an automobile body." The team found that when they entered this kind of high-resolution wind information—wind speed and direction for every 100 square meters of the landscape—the accuracy of even fairly crude models increased significantly. "Even using very simple algebraic models of fire spread, if you get better wind information, on average you get much more accurate predictions."

Curiously, Butler and his colleagues intentionally decided not to pay attention to other important weather factors like solar heating, surface type, and evaporation rates, and to focus instead just on modeling airflow. Maybe not surprisingly, their approach has been controversial for many in the meteorological community who are reluctant to throw out the complex physics that drive their large-scale weather forecast models. But by simplifying the approach, what Butler and his colleagues gained is a tool that can be used in the field.

"The decision space for fire managers is really narrow," says Butler. "They're making decisions several times a day, and for each one they've got only a few minutes, maybe an hour. For WindNinja to be useful, it had to be runnable by someone who uses it on a laptop once or twice a year, and it can provide a solution in less than an hour. That's why we simplified the physics."

Still, as great as such models may prove to be, most veteran firefighters think there's no substitute for on-the-fire-line experience. "I'm a technology geek," says Montana firefighter and incident commander Jon Trapp, who in an earlier career as a wolf biologist worked closely with other types of modeling techniques. He knows how much models can help. But then again, "If you've spent very much time as a firefighter, and you know something about the fuel, the weather, and the topography during a specific fire event, then you're going to be able to look up and pretty much know what's going to happen next."

AS SHOULD BE OBVIOUS BY NOW, THERE ARE FEW THINGS MORE important to predicting the behavior of wildfire than forecasting the weather. At the same time, there are few things more influenced by rapidly advancing climate science and computer technology. Such work is occurring both on a large scale, especially when it comes to making intelligent guesses as to where in the United States fires may start, and at a highly localized level, where forecasters try to pin down likely weather for a single drainage or mountain complex.

On the bigger stage, the National Oceanic and Atmospheric Administration, or NOAA, issues outlooks across the fire season from a storm prediction center in Norman, Oklahoma. Based on this information, regional NOAA stations can then issue one of two special fire warnings. The first is a fire weather watch, issued when strong winds and low humidity are likely to combine with lightning in the next twelve to ninety-six hours. The second is a red flag warning, issued when strong winds and low humidity, along with lightning, are likely to lead to wildfires within the next twenty-four hours.

Some of the most engaged, quick-thinking weather experts in the nation are incident meteorologists (IMETs), called in to the incident command center as part of a firefighting team. On arrival, these specialists will frequently visit the burn, often by means of overflights, to better clarify the inputs necessary for their predictions. There are eighty-five or so of these highly skilled, certified weather specialists in the United States. In the phenomenally busy firefighting season of 2015, IMETs were called in some 150 times. Yet the vast majority of wildfires in the West aren't staffed with an incident meteorologist. In those cases, wildland firefighters rely on the National Weather Service for localized forecasts. And while these are often highly accurate, it's extremely difficult to fully anticipate the size and movement—and the resulting changes in wind—of something as common as an isolated afternoon thunderstorm.

An incident meteorologist on a fire line brings with her a great deal of training in both large-scale weather systems—things like squall lines and thunderstorms—and microscale events like air turbulence and even dust storms. At the same time, she's trained in fire behavior and firefighting strategy, thus allowing her to converse well with the fire behavior analyst on the team, helping him anticipate what a given day of a wildfire might bring. Indeed, incident meteorologists are responsible for issuing painstaking reports of the most likely fire scenarios, offering those scenarios throughout the day and, depending on conditions, often throughout the night.

The IMET's spot predictions for areas as small as 75 acres depend heavily on information provided by the firefighters on the ground; armed with handheld Kestrel meters, they'll be describing to her the slope of the land, the air temperature, the surface wind speed and direction, the humidity, and the density of the trees. Meanwhile, the predictions she's crafting are coming across the firefighters' radios every hour, like clockwork. And what those firefighters are listening especially closely for is any report that tells them temperatures have warmed in the past hour while humidity levels have fallen.

The barometers and temperature gauges used in the recent past have mostly been replaced by a virtual world of information the incident meteorologist can access anytime, anyplace, with a satellite-connected laptop—linking together important data from a variety of sources: wind speed and wind direction maps, barometric pressure plats, even water vapor movement simulations. At any time the IMET can tune into dozens of National Weather Service centers, accessing Doppler weather, satellite images, and computer forecasting models—all part of what's known as the All Hazards Meteorological Response System, or AMRS.

She can also drill in on a very local level, drawing information from solar-powered units of the Remote Automatic Weather Stations, or RAWS, system. These local RAWS units are part of a network of more than two thousand such devices across the nation, maintained by various land management agencies, which measure temperature, relative humidity, and wind speed and direction. Most also contain a "fuel stick" gauge, which measures the internal temperature and humidity of a wooden dowel, giving important

Units of the Remote Automatic Weather Stations, or RAWS, system can be placed by helicopter in strategic locations near a fire, to transmit weather data to the National Interagency Fire Center in Boise, Idaho, via satellite.

information about the flammability of fuels in the forest. These stations transmit weather data to the National Interagency Fire Center in Boise, Idaho, via a satellite orbiting 22,300 miles above the earth. RAWS units are also set up so that firefighters can punch in a special tone on their handheld radios to instantly retrieve critical weather information.

She also brings her own RAWS units. Placed by helicopter in strategic locations near the fire, they send new data every ten minutes—not just about air temperature, humidity, and fuel stick moisture, but also peak winds, fuel temperature, and solar radiation. She can set the station to alert her when any of a number of critical thresholds have been met; what's more, the whole data set can be retrieved through a single website, allowing the entire fire team to stay fully informed. She also makes daily observations of the fire by air. And every two or three days she'll release weather balloons to collect critical information about conditions directly above the fire.

LET'S IMAGINE THAT AT OUR HYPOTHETICAL FIRE IN SOUTHERN Colorado, where the winds are beginning to strengthen, the National Weather Service has issued a red flag warning for the area immediately surrounding the fire. Critical wind shifts, which often happen in the afternoon during unstable weather conditions, are also forecast. And it's those gusty winds that might accomplish what firefighters and area residents most fear—pushing the fire over ridges and down canyons toward clusters of homes. In that scenario, the coming weeks will change the lives of almost everyone who lives in that sweeping, rugged slice of Colorado, not to mention testing the stamina and strategy of hundreds of firefighters.

FIGHTING FIRE

Heroic effort and tragic loss

THESE DAYS, ALL ACROSS THE WEST, HUNDREDS OF PROFESSIONAL wildland firefighters—especially those trained to have incident command duties—are spending their summers with pagers clipped to the side of their waistbands. Radios, too, which their neighbors in local coffee shops and grocery stores have grown used to hearing squawking softly as trickles of local emergency calls come in—people falling, car accidents, lost hikers. And fire. Most days you can pick them out of a crowd because they're dressed in Nomex fire pants and fire boots, ready to go anytime they get the call.

Fighting a fire is a little like waging a war. And given the stakes involved in the growing challenges with wildfire in the western United States, a remarkable number of local and state units, as well as federal agencies like the Forest Service, the Bureau of Land Management, the Park Service, the National Weather Service, and the Bureau of Indian Affairs—are in on the effort. During today's longer fire seasons when fire risk in California, New Mexico, Texas, Arizona, Wyoming, South Dakota,

A fire crew in southern California sets backfires, trying to halt the advance of the Poomacha Fire. The fire began with a structure fire in late October 2007 and eventually consumed 49,410 acres, destroyed 138 homes, and injured fifteen firefighters.

and Colorado can be extremely high all at the same time, people at the National Interagency Fire Center in Boise, Idaho, work overtime. Charged with coordinating logistical support, they lay out plans for how to allocate limited federal resources—including interagency hotshot crews and smokejumpers, bulldozers and helicopters and DC-3 tanker planes—to meet what can shape up to be a staggering level of need. In some places, the U.S. Army will be called in. Even firefighters from Australia and New Zealand might board air transports and make their way to fires in Colorado, Idaho, and Montana.

There are fifteen fire caches across the United States; the facility at the National Interagency Fire Center in Boise stores enough supplies to equip eight thousand firefighters in the field. Still, even with that kind of help, staging resources can be a difficult balancing act. While there's growing interest in "ordering the world," as some incident commanders call it, which means bringing in everything you can to a wildfire, such an approach can sometimes mean that smaller fires may as a result be short on resources and get out of control, with resources to fight them elsewhere, far away.

That's where the Incident Command System (ICS), a highly organized planning tool, comes in. Based in part on the command system employed by the U.S. Navy, ICS is a flexible, hierarchical management strategy—one used for every kind of emergency response you can think of, whether floods or terrorist attacks. The ICS was first turned into a potent emergency response tool in the 1970s in the wake of a series of wildfires in California that led to massive property damage, personal injury, and death.

As these early ICS teams began poring through the facts of what had happened in those devastating California wildfires, it became apparent that most of the mistakes came not from ill-advised firefighting techniques but rather from massive failures in planning, communication, and general coordination of resources. In part the problem was that each management agency—be it state, local, or federal—had its own set of guidelines and procedures for dealing with emergencies. When it came time to work together, it was like asking a group of people to do a complicated task without the benefit of a common language. As a result, the teams came up with what's become a superbly effective national emergency management strategy. (Indeed, in order to receive federal funds today, state and local governments must utilize it.) It's been used for such disasters as the World Trade Center bombing and Hurricane Katrina—and for thousands and thousands of wildfires.

BUILDING ON A CAREFULLY SCRIPTED RESPONSE PLAN, THE LOCAL fire management team assigns every wildfire in the United States a type number ranging from 1 to 5, with 5 being the least threatening and 1 the most severe. The type of fire an incident commander is certified for depends on how many years of study and testing he or she has undergone. The incident commander for a given fire—who may also be a firefighter with the local fire department—receives a page from the agency to which the fire was first reported, makes a quick call to check in with the communication center, and is quickly on the way to the fire.

"Even before we arrive on the scene," explains fire officer Jon Trapp of the Red Lodge, Montana, Fire Rescue Department, "we're doing detective work. If the smoke column I see in the distance is dark colored, that means the fire is burning in heavy timber. And if the column is going straight up, it means unstable atmosphere—conditions hard to predict." A more stable, and therefore less worrisome atmosphere, says Trapp, would show itself as a column going up and stopping, like an inversion; a column going up and laying over, on the other hand, would means winds aloft, which have the potential to affect air operations.

As the incident commander on our imaginary fire in Colorado is driving to the trailhead nearest the burn, on a jeep road ascending steeply through a mix of old ponderosa pine dappled with pockets of Douglas-fir and Engelmann spruce, he already knows fuel moisture is around 10 percent—dangerously low. The air temperature is 89 degrees F, which means the fire will be able to grow before the lower temperatures and higher humidity of nighttime hours. He knows the forecast, too, which calls for clear skies and strong southwest winds. And because this is his backyard, he knows the terrain where the fire is burning, including the general steepness of the slopes, the soil types, and what direction they face. (If need be, he can also consult thousands of topographic maps via his computer.) He knows the fire is burning in an area knee deep in downed timber—the result of a wind event that, as often happens, toppled dozens of acres of mature, insect-killed pine. And finally, he has a good idea of wind speed and direction—the latter factors determined by the movement of treetops. With a check of the sky the incident commander notes a line of cumulus clouds to the west—yet another sign that errant, gusty winds may be on the way. On parking at the trailhead, he straps a Pulaski to the outside of his "ready pack," lifts it to his shoulders, and begins the steep climb up a dusty trail, heading for the fire.

Clipped to his relatively small pack is his radio, the critical tool for communication, which he'll use to get progress reports from hand crews and smokejumpers, to direct aircraft operations, to report injuries, and to request resources. Inside the pack is a Kestrel meter, used for quick weather checks in the field. There's also a first aid kit, headlamp, bug spray, energy bars, water, a Leatherman multi-plier tool, a signal mirror, and marker panels to signal helicopters and other aircraft. It holds a spare shirt and jacket (made of fire-resistant Nomex, which can withstand temperatures of 900 degrees F without melting), dry socks, and a carbon face filter to breathe through if the smoke is bad (some firefighters just use wet handkerchiefs).

The pack also carries special flagging to mark dangerous standing burned trees. This latter item is more important than it might at first seem. A great many firefighters have been killed during mop-up operations, often from falling trees; indeed, more than 8 percent of firefighting fatalities are caused by these sorts of unexpected events, along with things like rolling rocks, downed power lines, and lightning strikes. And then there's the fire shelter. Measuring just slightly larger than a person, the fire shelter is the last line of defense for a firefighter caught in an emergency situation, when flames are bearing down and there's no possible escape. It's constructed of an outer layer of aluminum foil next to a weave of silica—a design that can reflect up to 95 percent of the heat from a nearby burn. Under that layer,

---

Firefighters, much like race car drivers, wear highly durable Nomex clothing for protection from intense heat.

and providing still more protection, is a thin chamber of air, followed by an inner layer of foil that's been laminated to fiberglass.

Fire shelters open very quickly—essentially a matter of grabbing two plastic handles and giving it a brisk shake. Once inside, firefighters loop their arms through hold-down straps that let them keep the shelter in place against the harrowing wind blasts that accompany big burns. The face is pressed against the ground, and the breathing kept shallow in order to avoid hot gases that can damage the lungs. Fire shelters can protect firefighters in temperatures of up to about 500 degrees F and have saved some three hundred lives, according to the National Interagency Fire Center, along with preventing serious burns to as many firefighters.

But the hope is that the fire shelter in the firefighter's pack will never have to be deployed—because safety is always, always the bottom line in firefighting. To this end, inside nearly every fire pack, in the pocket of nearly every pair of fire pants, tacked to the wall in the command center and taped to countless fire lockers and offices back home, is a card listing the ten Standard Firefighting Orders and the Eighteen Watchout Situations. Most wildland firefighters can all but quote them in their sleep. The orders—which aren't too different from the Standard Orders list of the armed forces—were developed by the Forest Service in 1957, based on a thorough review of hard lessons learned from sixteen deadly fires between 1937 and 1956. The Eighteen Watchout Situations—things like attempting a frontal assault on the fire and taking a nap near the fire line—were added years later in response to various firefighter fatalities.

BESIDES HAVING ALL THAT GEAR IN THEIR PACKS, EVERY WILDLAND firefighter in America has a simple yet terrifically useful resource tucked into his or her shirt pocket—a small spiral-bound notebook called the Incident Response Pocket Guide (IRPG). At just over a hundred small pages,

TEN STANDARD FIREFIGHTING ORDERS

FIRE BEHAVIOR

1. Keep informed on fire weather conditions and forecasts.
2. Know what your fire is doing at all times.
3. Base all actions on current and expected behavior of the fire.

FIRELINE SAFETY

4. Identify escape routes and safety zones and make them known.
5. Post lookouts when there is possible danger.
6. Be alert. Keep calm. Think clearly. Act decisively.

ORGANIZATIONAL CONTROL

7. Maintain prompt communications with your forces, your supervisor, and adjoining forces.
8. Give clear instructions and be sure they are understood.
9. Maintain control of your forces at all times.

If 1–9 are considered, then . . .

10. Fight fire aggressively, having provided for safety first.

EIGHTEEN WATCHOUT SITUATIONS

1. Fire not scouted and sized up.
2. In country not seen in daylight.
3. Safety zones and escape routes not identified.
4. Unfamiliar with weather and local factors influencing fire behavior.
5. Uninformed on strategy, tactics, and hazards.
6. Instructions and assignments not clear.
7. No communication link with crew members or supervisor.
8. Constructing line without safe anchor point.
9. Building fireline downhill with fire below.
10. Attempting frontal assault on fire.
11. Unburned fuel between you and fire.
12. Cannot see main fire; not in contact with someone who can.
13. On a hillside where rolling material can ignite fuel below.
14. Weather becoming hotter and drier.
15. Wind increases and/or changes direction.
16. Getting frequent spot fires across line.
17. Terrain and fuels make escape to safety zones difficult.
18. Taking a nap near fireline.

it's an amazingly well-considered set of checklists and assessment tools, all meant to guide firefighters in determining the severity of any wildfire they encounter and in beginning to initiate the most appropriate response. Using what's known as the Haines Index, for example, they can figure the likelihood that dry, unstable air will contribute to catastrophic fire spread; still another allows them to calculate risk according to lightning activity level.

For a Type 5 fire—the least severe type—more often than not the initial responder goes to the burn and from there perhaps calls in some local resources but often knocks down the fire with a small team by nightfall. But the blaze in Colorado is no Type 5 burn. Using his IRPG, along with his own experience and training, the incident commander quickly realizes that this is a Type 3 situation. He makes a call to the regional fire manager, requesting five engines, a smokejumper crew, two hand crews, a fixed-wing aircraft with retardant, and a helicopter. The goal is total suppression, and as quickly as possible.

What all this means to the people who live at the mouths of those canyons, and along the foothills of this rugged slice of the Rockies, is that they're about to see the delivery of a fully fledged organizational team—a highly trained "essential human infrastructure," as it is known. Within forty-eight hours an even higher-level incident commander will arrive to take charge of the operation, as well as a logistics officer. There'll be a camp officer, too, responsible not only for catering three meals a day and snacks but also for providing shelters, showers, and bathroom facilities for the firefighters. And there's a planning section chief, with direct ties to what's often a nationwide ordering of resources; with just a phone call or two she'll be able to arrange for everything from hotshot crews to helicopters to slurry bombers. Still, getting the needed personnel may prove to be a challenge as fires elsewhere compete.

A few members of this leadership team—an operations chief and a Type 2 incident commander—arrive right away as part of a predeployment team.

A dining hall feeds fire crews on the 2010 Eagle Rock Fire in the Kaibab National Forest in Arizona, which was caused by lightning and consumed 3,420 acres. Wildland firefighters can burn a whopping 6,000 calories a day.

Before they even land in the nearby town they swing into action, reviewing weather forecasts and fuel conditions and terrain types. At the same time, they check on what other fires are burning in the area, since this will affect the resources available to them, whether planes or bulldozers or hand line crews. Once they arrive on the scene, the person they'll be looking for is the incident commander who first arrived on the fire. "They'll want to know right away what's already been done," says firefighter Jon Trapp. "And what needs to be done next—get a clear sense of his biggest concerns. Together the team will do an immediate assessment of the resources already in place, as well as which ones are missing."

Then there's a long list of questions to consider. Where will briefings take place? Who are the liaisons from the sheriff's office and the police department (critical for traffic control, as well as in the case of evacuations)?

How many and what types of engines and crews does the local fire department have available? Where's the best place to refill water buckets by helicopter? What sort of remote weather data stations are in place in the surrounding forest? What if there's a need for a medical evacuation (medevac)? Who has bulldozers?

From that day on, at the start of every roughly twelve-hour operational period, the incident commander gives fire managers and division supervisors a comprehensive incident action plan, or IAP. Each plan clearly states what the priorities are, describes everyone's assignments for the day, and lines out who each person will be working with. The IAP also includes the most up-to-date weather forecast for the day, as well as safety and medevac instructions.

BASED ON THE INCIDENT COMMANDER'S EARLY REQUEST, FILED shortly after getting a good look at the fire, the fire team in southern Colorado is getting smokejumpers first thing. A dozen of these elite firefighters get airborne—on their way to Colorado from Idaho in a DC-3—within fifteen minutes of getting the call. In the years since smokejumpers were first on the job, in the early 1940s, about five thousand men and women have been trained and certified in this exhilarating, demanding precision work.

It was in the years following World War I that the Forest Service first started experimenting with the use of aircraft to detect fires in the western United States. By the mid-1930s they'd also begun trials (largely unsuccessful, as it turned out) of dropping water on burning forests. In 1939 attention turned to dropping men in parachutes—a system finally tested on the Nez Perce National Forest in Idaho in the summer of 1940, directed by the Ninemile Operations Center in northwest Montana. By the time the snows fell that year and ended the fire season, smokejumpers had dropped into nine wildfires, both out of the Ninemile base and from another in Winthrop,

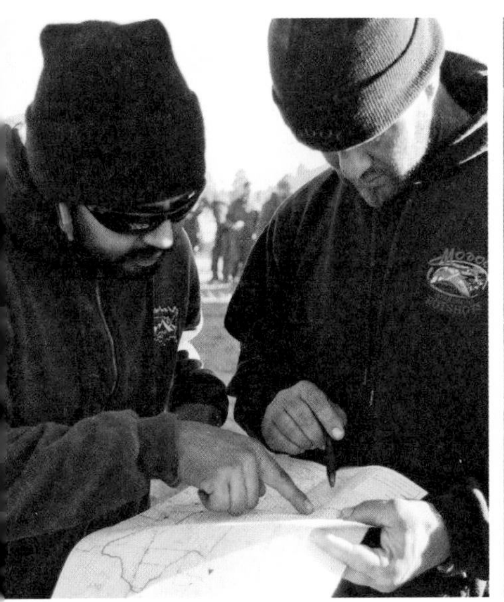

LEFT **Crew leaders from the Rocky Mountain National Park Alpine Hotshots and the Modoc Hotshots of northern California plan a day on a fire in Arizona's Kaibab National Forest.**

RIGHT **The McCall, Idaho, smokejumpers are flown to a fire.**

Washington, saving from loss an estimated $30,000 in timber, structures, and other resources.

Today, more than seventy-five years later, at the peak of any given fire season roughly five hundred smokejumpers will be on deck, waiting for the call. Getting to this point means about five weeks of rigorous training; indeed, only seasoned firefighters are even qualified to apply. If accepted, they will add to their already existing skills things like tree climbing and rappelling, negotiating obstacle courses, jumping out of a 40-foot-high "shock tower," helicopter and fixed-wing airplane use, as well as lots and lots of hiking with a 110-pound pack. And for some, plenty of time too on something called "The Mutilator"—an elevated pulley system used to haul jumpers into the air and then drop them about 30 feet, often with ground impacts of 15 miles per hour. It's a rigorous endeavor—so much so that attrition rates of 30 to 50 percent aren't uncommon.

Then there's the not-so-inconsequential challenge of parachuting itself—a version that in truth is a far cry from recreational parachuting. For starters, because fires often cause their own weather, including highly erratic winds, smokejumpers have to learn the fine art of dropping streamers out of the plane before they jump, using how those streamers behave to better anticipate what kind of turbulence they may have to deal with when making landings.

Imagine the spotter for a twelve-person crew called to our hypothetical Colorado fire staring with great intent out the door of a turbine-engine DC-3, gauging the burn below. Mostly he'll be looking for wind turbulence, using that information to choose a safe landing zone for the team about a quarter mile from the flank of the fire. Besides the hand tools they'll need to begin digging fire line, the dozen men and women will have enough food dropped with them to last a couple of days—just in case it takes a while for an air drop to resupply them. Their assignment will likely be to start digging fire line along the flank of the fire and then to use a drip torch to do a burnout, creating a blackened zone that the fire, with any luck, won't cross. Safety zones need to be identified. A lookout posted.

Given the fuel, weather, and location of our Colorado fire, a twenty-person hotshot crew—renowned for their highly elite physical abilities and firefighting skills—will also be on the way, hiking to the scene from a jeep road 4 miles from the blaze. The two teams, smokejumpers and hotshots, will be charged with working together to link their fire lines and thus keep the burn from dropping into drainages where people are living.

BEFORE CREWS ENGAGE A WILDFIRE, IT'S A HARD-AND-FAST RULE that four things must be in place: lookouts, charged with watching overall

A hotshot crew heads out with Pulaskis and chain saws to fight the 2014 Happy Camp Complex Fire in California's Klamath National Forest. So extreme were California's drought conditions at the time that six months after crews put out the blaze—after it had burned 134,000 acres—the fire rekindled.

fire behavior and communicating it to the team leader, the only exception being if the firefighters can actually see the fire; a communication link, to both the incident command center and members of the fire crew; an identified escape route in case the fire shifts and forces an evacuation of the area; and a safety zone, consisting either of already burned ground or a place with bare soil sufficiently far away from any burnable fuel.

In November 1956, eleven wildland firefighters were overrun by flames and died in San Diego Canyon on the Inaja Fire in California. In the aftermath of that tragedy, the Forest Service mandated that from then on, all firefighters would identify a safety zone—a place where it would be possible to survive without a fire shelter. Yet years later, even as recently as the early nineties, no one could really define what made a good safety zone. As one former firefighter put it, "What constituted a safety zone was totally up to each of us—based on experience, intuition, or guessing." But scientific research done since 2000 on so-called energy transport in wildfire—in other words, how heat is transferred in space and time—led to guidelines in place today about how far a firefighter needs to be from a flame to avoid burn injury.

Generally speaking—depending on vegetation—if the flame height is 20 feet, a three-person crew needs to claim a zone that keeps them 80 feet from the flames—which translates into an area of about a half acre, or roughly half the size of a football field. If the flames are 100 feet high, firefighters need to be 400 feet away, which equates to about 12 acres; 200-foot flames would require a safety zone of 46 acres. In addition, these safety zones can't be upwind or upslope from the fire. And because of the potential for high winds, great care also has to be taken in areas with narrow canyons and mountain saddles.

If any one of these safety measures—lookout, communication link, escape route, or safety zone—is missing, no one engages the fire. No exceptions.

THE DECISION TO CUT A FIRE LINE IS BASED ON CAREFUL, THOROUGH research and training. Using information from what's known as a fire behavior hauling chart, the firefighters who first show up on that southern Colorado fire know that because the flame height of the burn is only about 4 feet, their best approach is to attack the head and flanks of the fire with hand tools. But they also know that if those flames reach 6 to 8 feet, hand lines aren't enough. At that point, they'll be sent around the flank of the

A saw crew on the lightning-caused Whiskey Complex Fire south of Garden Valley, Idaho, cuts fire line in July 2014.

burn to cut fire lines in places that will help drive the wildfire away from subdivisions and water supplies.

Fire lines are literally lines in the forest or grass or scrubland, varying from as little as 8 to 24 inches wide—or if dug by bulldozer, 8 to 10 feet wide. To a large extent the width of the line cut by the crews depends on the size of the fuel next to it; if there are tall trees all around, a wider line will have to be cut. Either way, the idea is to clear away vegetation down to bare soil, and by doing so, to help contain an advancing blaze.

In forested areas, it works like this: at the head of the team is a saw crew—two to four firefighters working at a feverish pace, using either hand saws or chain saws to cut both live and downed trees. Behind them, in turn, is the rest of the hand crew. Those in the front carry a Pulaski, or P-tool, which they use to take quick swipes at debris, as well as to hack out roots. Next in line come team members armed with tools that have both a cutting edge and a straight edge, the latter resembling a very sharp hoe, which they use for cutting roots and dragging away litter. Finally, at end of line is a small group wielding McLeods, which have rakes on one side and flat edges on the other, for final scraping down to bare earth.

A fire line being cut by professional wildland firefighters is a thing to behold—a muscular yet in many ways graceful flurry of cutting and scraping and dragging away, trying whenever possible to link the line to existing natural features such as rock outcrops. The very best crews can construct lines through a forest incredibly fast, nearly at the pace of a person walking. This work involves an extraordinary level of physical activity—enough, in fact, that a member of a line crew can burn an astonishing 6,000 calories a day.

But wait a minute. Is any wildfire really going to have the slightest trouble crossing a foot-wide line of bare dirt? Or for that matter, even a line the width of a bulldozer blade? Not really. But cutting a fire line is just the first step. Next comes a burnout, which involves intentionally setting fire along the hand line, thereby consuming unburned fuel between that line and the fire front. These are typically set using a drip torch, which is a small handheld can of flammable fuel that pours through a nozzle across a flaming wick, thereby dropping burning liquid onto the ground. (On some occasions burnouts are started with hand-held flares called fusees.) The theory is that when this strip of fuel is burned in a controlled manner, the advancing fire won't have anything to burn when it reaches the line and thus will stop. If a wider burn line is required, another burn can be lit just above the initial one, and then another, and so on, each strip of burned ground linking to the one below it.

The other similar, though more complicated, tool in a firefighter's arsenal—complicated enough that it can be launched only by the operations

LEFT It's critical that sawyers on hand crews know how to keep their saws in perfect condition. Notably, it takes years of training and testing for a crew sawyer to gain certification as an advanced faller.

RIGHT A member of a Forest Service hand crew mops up using a McLeod.

section chief—is the backfire. This is also a fire set inside of the control lines established by hand crews, but in this case the action is meant to consume a considerable amount of fuel in the path of the burn. Because of their size, backfires are also used to change the direction of a wildfire's convection column, which in the process changes the direction of the fire itself. The reason this works is that the heat from a wildfire tends to pull a flow of air toward the base of the fire—in other words, feeding itself—which in the process draws the backfire in, essentially backing the wildfire onto itself.

As brilliant as this may sound, setting backfires is a highly technical and potentially very risky endeavor. It requires careful calculation based on lots of experience. For example, the backfire has to be far enough away from the main burn that it creates a dead zone of burned trees rather than simply adding strength to the existing blaze. And that means thinking carefully

A crew manages a burnout operation on a California wildfire. While most burnouts are started with drip torches, they can also be ignited with flare guns.

about what will be the likely speed and direction of movement of both the main burn and the backfire in the coming hours.

Meanwhile, as all this is going on, teams of firefighters known as holding crews have the job of making sure the fire line holds, which means putting out fires that suddenly erupt on the wrong side of the control line, ignited by flying sparks, pine cones, and embers.

The act of using fire to fight fire can be incredibly important to firefighters, since having access to a place where the fuel has already been burned away is a key to staying safe near fast-moving flames. In fact, if you were listening to radio transmissions on any of several hundred fires in the West, you'd hear hotshots and hand crews informing the communications center that they have "one foot in the black," meaning they're in a place where they can easily seek safety.

The Geronimo Hotshots set backfires in an effort to halt the advance of the lightning-caused Big Windy Complex Fire near Galice, Oregon, in August 2013.

NO WILDLAND FIREFIGHTERS ARE TOUGHER OR BETTER TRAINED than hotshots. These elite forces got their start on the Cleveland and Angeles national forests in southern California in the late 1940s, taking their name from the fact that they were routinely sent into the hottest part of a wildfire. There are about a hundred hotshot crews in the United States, and bad situations are their bread and butter.

The crew superintendent is responsible for directing her firefighting team on the ground in extremely dangerous conditions. The job she signed on for is to lead nineteen other people through some of the most intensely stressful, physically and mentally demanding situations imaginable. And for that she needs to be the sort of person who can remain calm in the midst of extraordinary pressure. The number one thing on her mind is to come home from each assignment with every one of her people alive and well.

In order to do the best job possible, beyond her time on the fire line she's been trained as a medical first responder, studied wildland fire behavior and ignition operations, and taken dozens of other training courses—on chain saws and air operations, portable pumps and water use, entrapment avoidance, firefighter health, hazard tree safety, and more. She's spent dozens of hours perusing the tactical decision games library at the base and has read countless pages analyzing incident reports and summaries of lessons learned on other fires. She's had to show physical strength through annual testing, as well as to keep her medical first responder certification up to date. She can run a mile and a half in slightly over eight minutes, and can do forty-five sit-ups or thirty push-ups or fifteen chin-ups in a single minute.

Matt Holmstrom, superintendent of the Lewis and Clark Hotshots in Great Falls, Montana, talks about the intense scrutiny hotshot leaders give to the fire on first arriving—gathering all the information they can, looking at every angle with their own eyes. "It's not that we don't trust the incident commanders—that's not the point. But it helps to put our own eyes on things, hold up what we know we can do against what we're being asked to do." What's more, Holmstrom says, it's important to keep in mind that the hotshot crew you bring in in June isn't the same crew that shows up in the fall. By late in the season a crew will likely have spent at least a hundred sixteen-hour days fighting wildfire—enough labor and mental intensity to leave some, as he puts it, "having gone a bit to rubber."

Even so, while the end of a long fire season may bring a certain physical weariness to wildland firefighters, what keeps their decisions on target is their wealth of experience in the face of demanding and rapidly changing conditions. Hotshots can also get a hand now and then from emerging technology. "Take Gaia GPS mapping software," says Holmstrom. "With that up and running on a laptop I can know what the next ridge looks like before I get there. That can be a big help. At the same time, I can look at

The Payson, Arizona, hotshot crew heads out to a fire line with fire hoes in hand.

weather radar right from the fire line and know right away what a nearby storm is doing."

For all that great technology at their fingertips, though, Holmstrom says firefighting research has lagged a bit when it comes to understanding what it is that leads to errors in human decisions. "Sometimes people can have all the right information but still make the wrong choice," he says. In his ongoing research, thus far he's identified eight human factors present in what are known as tragedy fires. They include things like having less than two years or more than thirteen years of experience; the time of day (2:00 p.m. to 6:00 p.m. are especially dangerous for fire entrapments); wearing too many hats, or poorly defined leadership; conflict on the fire line; and poor communication. By better understanding such factors, Holmstrom hopes to take longstanding safety research beyond studies of how fire behaves and into how humans get into trouble in the first place.

Every member of every hotshot crew in America knows that fire conditions in many parts of the West are becoming more severe, with temperatures rising and humidity and precipitation dropping, greatly increasing the flammability of the forest. They know too that storm energy is growing, giving rise to stronger, more unpredictable winds. All of which means that these wildland firefighters not only are busier but also increasingly need every tool and training available if they're to stay safe in this demanding work. They are also painfully aware of wildland firefighter fatalities; and it's not simply that they've heard about them, but that they've reviewed the investigation reports of those deaths with painstaking dedication. And of those fatalities, perhaps none in recent years has been more troubling to them than what happened in June 2013 in Arizona, at Yarnell Hill.

THERE, WHAT BEGAN INNOCUOUSLY AS A LIGHTNING-CAUSED blaze in a boulder field, posing no threats to people or structures, was met initially with a Type 4 response, which seemed adequate—that is, until erratic winds caused the fire to swell and prompted a request for a Type 2 incident commander twenty-four hours into the fire. Three hotshot crews were ordered up. The Granite Mountain Hotshots from Prescott, Arizona, arrived at the incident command post at eight the next morning, a day and a half after the fire started, and were told they had safety zones at the Boulder Springs Ranch as well as within several areas previously burned by the fire. By now the blaze was close to 500 acres.

Three lookouts were stationed on the fire by one o'clock that afternoon. Each established geographic trigger points for himself and his crew—places that if reached by the fire would cause an immediate evacuation to a safety

Wildfire is incredibly dynamic, requiring a steady flow of updates about current conditions, as well as an ability for team leaders to adjust tactics at a moment's notice. This hand crew on the Springs Fire in Idaho in August 2012 gets instructions from the crew boss.

zone. With a storm brewing, and with predictions of 45-mile-per-hour wind gusts, with the heat rising and humidity dropping, by three o'clock the district forester and the incident commander ran a complexity analysis and realized they needed a Type 1 management team—the highest level possible. Then, at 3:30, the winds changed course by 90 degrees. Spotting started to happen, which meant embers drifting across fire lines to start new burns behind the firefighters. It started looking like the fire might be heading straight for the town of Yarnell; and if it did, in those conditions it would probably take less than two hours to get there. The lookout for the Granite Mountain crew saw the fire hit the first trigger point he'd set up, and after informing the crew captain, he moved back to his predetermined safety zone. Based on that communication with his captain, he believed his crew was safe.

But on their way to a safety zone at Boulder Springs Ranch, the Granite Mountain Hotshots started down into a box canyon and lost sight of the fire. They were completely surprised when a dramatic shift of wind pushed it around a rocky ridge and right toward them at the terrifying rate of 10 to 12 miles per hour. With the flames only five minutes away, the crew set about desperately clearing away the juniper and chaparral with hand tools to prepare a place to deploy their fire shelters. Because these shelters are not designed to withstand sustained heat, if firefighters haven't sufficiently cleared fuels around their so-called deployment site, they can die. Which is exactly what happened on that tragic day, when nineteen of America's most elite firefighters perished inside their shelters—marking the largest single firefighter catastrophe in wildland firefighting history.

The investigation that followed showed that the blaze had totally surpassed the incident command team's expectations, leaving them scrambling in the face of a growing set of challenges. The Arizona State Forestry Division—the agency in charge of fighting the fire—had failed to follow their own checklists and attack guidelines, in an important cautionary tale for firefighters across the country.

AS WILDFIRE PROLIFERATES IN THE AMERICAN WEST, SO DOES THE use of every available resource, new and old—including a wide range of aircraft. Indeed, by now the Forest Service all by itself uses a fleet of around a thousand aircraft every wildfire season, some of which it owns, others secured by rental or contract—often at a cost of more than $200 million annually. Yet because of how active fire seasons have become, there can be shortages of almost everything. Some of the airplanes and helicopters sent to big burns are brought in by the Air Force.

Fixed-wing craft are used to survey a burn, giving managers a bird's-eye view of the action; others transport fire crews, including smokejumpers—a job that also routinely gets assigned to helicopters. As for the rest of the fleet, their work consists mostly of fighting the fire—dropping water or retardants to slow the advance of the flames so that the people who have boots on the ground have a chance to get the upper hand. Retardant is often the first order of business, applied to slow the fire, after which precision water drops are made onto the flames. Where there's much concern about a fire moving into a canyon system with homes at the mouth, the order will be given to "paint the ridge" with retardant in an effort to keep the burn from crossing over.

Air tanker firefighting operations started out in the most humble fashion—a Ford Trimotor plane in 1930 dropping a wooden beer keg full of water on a remote burn. By the mid-1930s, large military bombs full of water were being fixed with so-called proximity fuses, which were ignited just above the target to saturate the flames. Other bombs, meanwhile, were intended to shatter on impact, tossing water or chemicals into the path of the advancing fire.

But after World War II things changed quickly, not only in the type of aircraft used but also in what got dropped on the fire. To be effective, water has to be dropped in exactly the right location—a task that can prove remarkably difficult in gusty winds. So in the mid-1950s the Forest Service started mixing sodium calcium borate into water, spreading it ahead of the flames to interrupt the chemical process of the burn. The mixture had the unfortunate side effect of sterilizing the soil—hardly a good thing, considering that regrowing vegetative cover is essential to preventing erosion. The retardants now being dropped, by comparison, contain fertilizers, colored red or pink so that the pilot can "join lines" on multiple drops.

The smallest of the aircraft working fires today is a single-engine air tanker (SEAT), with a capacity of about 800 gallons of water or retardant. Then there is an amphibious plane called a super scooper, capable of scooping water on the fly from area reservoirs. There's also the Erickson

A "super scooper" plane takes on water to drop on flames raging in 2014 in Washington's Carlton Complex Fire, which scorched 256,108 acres, the largest wildfire in Washington history.

S-64 Aircrane, equipped with a sea snorkel, capable of filling a 2,600-gallon tank while hovering above a river or lake. But again, such resources might be in short supply, being called on to fight fires from Alaska to New Mexico.

As a burn increases in size—moving faster, the walls of flame growing taller—still larger aircraft are brought on. Among the most familiar is one that's been around a long while—the DC-10, which started as a commercial airliner way back in the 1960s. Carrying a whopping 11,600 gallons of water or retardant per load, four times the amount carried by almost any other tanker, it's an extraordinary sight to see one rolling overhead and then unloading a 3,000-foot-long stream of red-dyed retardant, often while flying just 150 or 200 feet above rugged terrain. The DC-10 has a surprisingly narrow turning radius; what's more, it can maneuver especially well, including

The 2015 Cable Crossing Fire in southern Oregon was attacked with water and retardant drops soon after it started; nonetheless, by the end of the week it had grown to 1,500 acres.

making steep climbs, though in part this is because it carries far less fuel on fire runs than it would were it doing work as a commercial airliner.

Increasingly, large tankers like the DC-10 are being employed even on modest wildfires as first-strike tools—an approach that will probably become more common as drought continues to unfold, especially with fires that occur in the wildland-urban interface. There are even plans to convert a bigger-than-usual version of the already giant Boeing 747 into a firefighting tanker—creating a plane that could haul more than 20,000 gallons of retardant.

Like so much of firefighting work, air tankers are effective because of a team that extends well beyond the pilots themselves. It's firefighters on the

A Douglas DC-7 (a smaller forerunner of the DC-10) drops fire retardant on the Government Flats Complex Fire near The Dalles, Oregon, in 2013.

ground who call in retardant drops—giving location information not to air tanker pilots directly but to small lead planes that in turn direct the big tankers to just the right place. (Planes are always directed away from the main smoke column, by the way, as those columns usually contain extremely turbulent winds.) The tankers can either let loose their cargo all at once, in a line up to 50 feet wide, or in pulses along different points on the fire line.

While planes are critical for dropping fire retardant, as well as for doing aerial surveys, helicopters are no less valued—for their ability to both transport fire personnel and make extremely precise water drops on an active fire. For this latter task the copter will be outfitted with helibuckets (one of which is a very popular model known as a Bambi bucket)—receptacles that can hold between 72 and nearly 2,000 gallons of water, hung by cable from the bottom of a helicopter specifically chosen to handle the weight of

Near Bastrop, Texas, a crew attaches an aerial water delivery container called a Bambi bucket to a CH-47 Chinook helicopter.

the water, and operated by the crew with a release pull. Helibuckets can be refilled from nearby reservoirs, or even streams and rivers—some as shallow as a foot and a half—thereby saving precious time by avoiding trips back and forth to the base.

BACK AT THE FIREFIGHTING COMMAND POST IN SOUTHERN COLOrado, set up in the elementary school, things are happening at a rapid pace. The effort now includes two hundred firefighters, three bulldozers on lease from a local contractor for cutting fire lines, three planes dropping fire retardant ahead of the flames and doing precision water drops, two fixed-wing survey and pilot planes, and four helicopters.

As the fire advances, a structural specialist is brought in to advise the

The CH-47 Chinook drops water from the Bambi bucket onto a blaze outside a home threatened by the Bastrop County Complex Fire, which in September and October 2011 destroyed 1,673 homes and killed two people.

incident commander about what types of breaks might be created to protect the down canyon homes. But that job all by itself is a tough one. The walls are steep, littered with downed lodgepole pine. Bulldozers can't get in to build breaks in that kind of terrain, and hand crews will need more time than they would in cleaner, less cluttered conditions. And if the fire spots, fire line will be needed there, too.

Suppressing the fire is getting more and more complicated. By running a careful analysis that takes into account things like the most recent forecast of the incident meteorologist, the increasing threat to homes and other structures, the difficulty of the terrain and the blown-down timber, as well as the types of resources available, the incident commander and his team realize that they need to make a call they were hoping to avoid. It's time to evacuate. And so later that afternoon comes the knocking on the doors of

In wildfires, homes aren't typically consumed by great walls of flame but rather set on fire by embers launched vertically from the main blaze. The 2011 Wallow Fire in Arizona destroyed thirty-two homes and damaged five others, including the one pictured here.

hundreds of residents, a task carried out by local law enforcement—letting everyone know that the time has come to leave.

It's easy to imagine that in wildfire tragedies, homes and barns and businesses are consumed by great walls of flame. But that's rarely how it happens. Most structure fires are actually caused by embers or firebrands (which include things like flaming pine cones) launching vertically from the main blaze and traveling considerable distances, pushed by the wind stream for a mile or two or even three, then landing in people's yards or on the rooftops of their homes.

It's important to note that with fire literally dropping from the sky onto

Fire crews do structure protection on the Carlton Complex Fire in Washington, which consumed around three hundred homes. While it's often possible to keep homes from igniting, if they do catch fire, land management agencies are generally not equipped to fight the blaze.

people's homes, it's not the land management agencies like the Forest Service that respond, but rather local fire departments, often staffed with volunteers. While the Forest Service, the Bureau of Land Management, the Bureau of Indian Affairs, and state departments of resources can and do try their level best to keep wildfire away from subdivisions, they simply don't have the bunker gear, breathing apparatus, or other equipment needed for structure protection. For example, the water flow rate alone needed to fight a house fire far exceeds what most agency-owned wildland firefighting engines are capable of.

This is why when it comes to requesting resources to fight wildfire, the appeals are often for structure protection equipment and personnel. In busy fire seasons, and there are a lot of those these days, fire engines may be in

Washington governor Jay Inslee offers a heartfelt thanks to firefighters working the Chelan Complex Fire in the summer of 2015.

such short supply that they have to be shuttled from thousands of miles away. "In my town of Red Lodge, Montana," says firefighter and wildfire incident commander Jon Trapp, "our local fire department can deal with one structure fire really well—and two structure fires fairly well. But we can't deal with four or more."

And that can be a horrific problem. When firebrands ignite a house in a subdivision, if the homes are built within 30 feet of one another they're very likely to end up burning in a kind of domino effect, igniting by radiant heat one after the other, every fifteen to twenty minutes. Furthermore, burning structures also tend to give off hot embers, which can also spread the fire by traveling through the air and landing on other surfaces far away.

But let's say the people whose homes are threatened by our imaginary Colorado blaze get lucky. After they're evacuated, with flames advancing

A group of military veterans, now serving as specially trained firefighters for the Bureau of Land Management, mops up at the end of a California forest fire.

quickly and firefighters talking about doing “ring ignitions” around homes in the middle of the night to burn away fuels from the structures, let’s imagine the wind shifts, turning the fire back on itself, and an early autumn rainstorm settles everything down long enough for containment of the blaze. The beginning of the end of that wildfire could easily come a month or more after it begins—in part, no doubt, thanks to several million dollars’ worth of firefighting effort.

When it comes to the big fires happening today, in a lot of cases what puts them out finally and fully is major rainstorms, or just as often, the snows of autumn. But even when firefighters are not able to stop the direct advance of a burn or to extinguish it completely, they often do a brilliant job of essentially herding it away from subdivisions and waterways by changing its angle. Mop-up operations can continue for weeks, even months. And after mop-up is done and the firefighters have gone home, many residents who have been evacuated from their homes still dream of fire. Still feel a lurch in their stomachs at the mere smell of smoke.

AFTERMATH

Nature in the wake of the burn

IN THE AFTERMATH OF A WILDFIRE, BESIDES THE CHARRED ACRE-age and blackened forests, watersheds may have been severely damaged. When it rains in the fall or the following spring, compromised soils may run off and fill the local water supply with sediment. In addition, many slopes will have been greatly destabilized by the burning away of the vegetation that once bound the soil and rocks, which could well lead to debris slides damaging roads or even homes located at the mouths of steep canyons. Trails will have been lost, and with the Forest Service's greatly diminished recreational budget, a lot of them are unlikely to come back again. And there are economic consequences, too. Some locals who serve tourists will lose so much money that they'll be forced to hang it up, sell out, and move away. And though the trees will be back—the first seedlings up and reaching for the sun by late the next spring—none of the residents of the burned area will live long enough to see the return of a mature forest.

In the natural world, as in human communities, the impacts of wildfire

The beautiful, aptly named fireweed was among the first plants to rise after the 1988 Yellowstone fires. Fireweed is well known in many parts of the world for its ability to quickly colonize after fire. It was not only among the first plants to appear after the eruption of Mount Saint Helens in 1980 but also colonized burned ground after the bombing of London in World War II.

go on long after the flames go out. In fact, many of the ecological processes that wildfire gives rise to continue on for decades. Some of these processes are quiet, barely noticeable, while others, including erosion and landslides, can be hugely dramatic. Sometimes just months after a fire goes out, grasses and sprouts from smaller plants like currant and fireweed are already showing up, and even the sprouts of trees, including lodgepole and aspen. And

LEFT Many wildfires leave the ground significantly more nutrient rich than it was before the burn. These plants began sprouting two months after a wildfire.

RIGHT Just one year after the 1988 Yellowstone fires, much of the land was again rich with grasses and wildflowers.

if nothing happens to disturb the process, those pioneer tree species may go on for eighty or a hundred years, until a wave of shade-tolerant trees like spruce and fir spring up to claim their own day in the sun. But make no mistake about it. Sooner or later it will all burn again.

AT ITS MOST ESSENTIAL LEVEL, WILDFIRE IS A FORCE THAT REARranges energy. When plants and trees burn, the energy they hold within their biomass gets released as heat and gases and charcoal and nutrient-rich ash. A whole host of wildflowers show up in recently burned locations, including paintbrush, lupine, and the aptly named fireweed, as well as a

Some pines, including the lodgepole, have developed serotinous cones that stay closed on the tree and open only when exposed to flame.

wide range of grasses. Many of these plants do extremely well after a burn—not only because competition is down but also because the ash left in the wake of a wildfire contains essential nutrients that were previously held in the biomass of the trees. After the Yellowstone fires, elk especially benefited, enjoying more, better quality graze than they had had before the fires.

Wildfire also triggers some conifers—such as the giant sequoia, jack pine, and lodgepole pine—to reseed. Their cones stay closed on the tree and open only in response to heat. (In general, cones that open only in response to some environmental trigger are termed serotinous cones.) Not by accident, the time necessary for the cones to open on a lodgepole in the presence of flames is about twenty seconds—almost exactly the time it takes for a tree to "crown out," or burn from base to tip. In this way the burned trees seed the next generation of conifers, which in its own time will—if not disturbed by other agents, like insects or disease—burn and reseed again.

LEFT When lodgepole cones do open, they quickly seed the next generation of forest.

RIGHT Just three months following a burn, eucalyptus trees are already showing new life, sending out branches from specialized buds under the bark.

Elsewhere in the plant kingdom, if fire destroys the branches of a eucalyptus, the tree will sprout new branches from specialized buds located under the bark of the trunk. Both aspen and scrub oak will resprout from the roots in the aftermath of fire burning the major portion of the tree. Still other trees, including ponderosa, larch, and giant sequoia, have thick fire-retardant bark. The genus of flowers aptly named fire lily is not merely prone to blooming only after wildfire but does so incredibly fast; in fact, one species reaches full flower in just nine days in the vegetation-free landscapes that follow a wildfire.

But burned landscapes can also provide a toehold for invasive species—plants that may use more water, be less nutritious to wildlife, and by virtue of their density and growth patterns encourage the spread of fires in the future. For instance, cheatgrass, probably originally from Eurasia, not only moves in quickly following a burn but because it completes its life cycle by early summer, tends to provide enormous amounts of dry detritus during

Highly invasive cheatgrass is taking over much of the West. In some places it achieves growth densities of twelve thousand plants per square yard.

fire season—creating much hotter burns than would otherwise occur. Hot enough, in fact, that the more nutritious native plants, some of which are often more heat sensitive, can't survive; and that, in turn, allows cheatgrass to gain even greater dominance. This leads to a cycle described by some land managers as "quickly burn and quickly return." At this point, in the Great Basin alone (a 209,000-square-mile area that encompasses most of Nevada, half of Utah, and portions of Oregon, California, Idaho, and Wyoming), cheatgrass is the dominant ground plant across some 25 million acres.

Even under better circumstances, with lower-temperature fires, native species may lose out. While the plant succession that starts up shortly after a fire favors native grasses and flowering plants, cheatgrass is simply better at spreading quickly. If in the process it uses up most of the nitrogen and phosphorus stored in the soil, native plants are squeezed out.

Often on the heels of a cheatgrass invasion comes another invasive known as medusahead, which can grow at densities of more than a thousand plants per square foot. This low-quality forage grass now occupies considerably more than 2 million acres of western lands. Notably, medusahead is high in silica, which greatly slows the decomposition of dead plants; all of which means that for years after they die, the stems and dried leaves remain available as tinder for wildfire.

Finally, sometimes these invasive species are able to colonize more quickly due to unintended consequences from past fire management mistakes. For example, our overly aggressive fire suppression efforts have allowed western juniper, a native tree, to greatly expand its former range, in the process displacing other native plants. When conditions are right for a juniper forest to burn, it often puts out incredible levels of heat, consuming everything around it. And it's exactly in the wake of those sorts of burns that plants like cheatgrass, medusahead, and buffelgrass can quickly move in and thrive—again leading to greater fire danger. In 2006 a series of fires burned nearly a million acres in the Chihuahuan Desert of west Texas, an event that in large part was fueled by invasive species. Likewise, in 2005 wildfires burned nearly 250,000 acres of Arizona's Sonoran Desert, with invasive plants fueling fires in places where fire had never before been recorded.

WILDFIRES DON'T JUST AFFECT THE THINGS THEY BURN—TREES, grasses, shrubs, and ground plants. They also impact the myriad creatures that make homes in those landscapes. It's easy to imagine deer and elk, mountain lions and coyotes fleeing in terror before towering walls of flame. But that really doesn't happen. In the closely observed Yellowstone burns of 1988, while large animals did move along in front of the flames, they didn't flee in panic like Bambi did in the Disney movie of the same name. Indeed, elk and bison were seen grazing, even lying down resting, just a couple hundred yards from the flames.

In possibly the most famous wildfire photograph ever taken, elk observe a wildfire in Montana's Bitterroot National Forest. Most animals don't run from walls of flame, as old Disney movies might have us believe.

All in all, casualties tend to be light—a circumstance that has much to do with the fact that creatures of the forest have been evolving alongside wildfire for a very long time. Of the tens of thousands of large animals that reside in Yellowstone, only 259 are thought to have been killed by the 1988 fires: 244 elk, 9 bison, 4 mule deer, and 2 moose. When researchers examined the carcasses of 31 of those elk to determine cause of death, 26 had coatings of soot in their throats below the level of their vocal cords, which means more than likely they died of smoke inhalation. Shortly after the fires went out, the carcasses of these large mammals were seen being fed on by grizzly and black bears, coyotes, golden eagles, and ravens.

Red squirrels may disappear for two or three decades following a major fire, since the conifer seed cones they live on will have been destroyed. Yet other rodents (mice, rabbits, and ground squirrels, all of which eat grass and

A Yellowstone elk wanders through a just-burned landscape in Yellowstone, 1988. Many of the park's burned lands quickly gave rise to rich mats of grass, which in turn provided highly nutritious graze for resident ungulates.

low-level plants) will often thrive, as will many of the raptors that eat those species, including hawks and owls.

Creatures with less mobility than large and small mammals, including reptiles and amphibians, tend to react by burrowing into the soil. Some use the advancing flames of a burn to their advantage, nabbing insects or snakes moving overland in front of the burn. Certain beetles equipped with infrared sensing organs can detect low-grade burns on logs, where fleeing insects provide a ready food supply. In places with shallow wetlands, amphibians often benefit from fire—at least the more modest fires that were once typical—because such burns keep woody vegetation from eliminating sunshine and open water.

Flickers and woodpeckers, which forage for insects under tree bark, usually have an easier time after a fire, because when sap burns off a tree it can leave those insects easier to reach. Other bird populations may initially decline in the wake of a burn but be doing much better a decade or so later than they were before the fire happened. Studies in southwest Oregon following the Quartz Fire of 2001, for example, showed that populations of the increasingly rare olive-sided flycatcher went down fairly dramatically following the fire; yet as the years passed, flycatcher numbers significantly increased thanks to the fact that standing dead trees left in the wake of the fire provided good nesting sites for the birds, and that new shrub growth led to more insects and therefore more food. So too did conditions improve with time for goldfinches, house wrens, and Lazuli buntings. As a rule, in the decade following a fire there may be a shift in bird species based on feeding habits: those birds that tend to search tree foliage for food have the greatest presence before the fire, while ground brush foragers do better after the burn.

If we can say with certainty that many species benefit from regular stand-maintenance burns, it's less clear what the effect will be of today's landscape-altering megafires. Such big fires can have a dramatic effect on

A western pond turtle survived the 2013 Rim Fire in California's Stanislaus National Forest by burrowing into the soil under a log.

soil erosion, for example, which can lead to heavy sedimentation of area streams—an effect that can last for years, having significant impact on native fish. Additionally, many native fish species succeed within a very narrow range of temperatures; more direct sunlight on the stream bottom can easily warm the water beyond what they can handle. Such big fires can also reduce the amount of water that soaks into the natural aquifers by increasing the amount of sunlight that hits the snowpack and causing it to evaporate before it can permeate the ground. This means less water will be available to creeks and wetlands through the warm months of summer.

Big fires, coupled with climate change, are also shifting the mix and structure of vegetation. This means a number of habitat-sensitive animals—the regal moose and the secretive pine marten being good examples—could in the decades to come suffer significant population declines.

A LESS FREQUENTLY DISCUSSED PHENOMENON IN THE AFTERMATH of wildfire involves large debris flows, generally consisting of a saturated slurry of rocks, fine-grained particles, and woody material. These frightening events tend to occur on steep slopes that have been destabilized by wildfire burning off the vegetative cover, then later saturated by heavy rains. Such conditions aren't hard to find in the West, from Big Sur to Lake Isabella, California, from Mount Rainier to the Front Range of Colorado. What's more, the problem is likely to grow worse with megafires in the picture, given that such burns can sterilize soil, thus delaying the growth of new vegetative cover; that, in turn, creates perfect conditions for debris flows and landslides to occur when big rains begin to fall.

Such events can be extraordinarily dangerous. A debris flow following Colorado's South Canyon Fire in 1994 barreled across all four lanes of Interstate 70 near Glenwood Springs, trapping thirty cars and sweeping two people into the Colorado River. In northern Utah between 2000 and 2004, twenty-six debris flows occurred following seven wildfire events. In the fall of 2003 in southern California, two wildfires—one called simply Old, and the other Grand Prix—burned the vegetative cover from the mountainsides near Cajon Pass, located just above the town of Devore. Later in the year, beginning on December 24, more than 4 inches of rain fell on those denuded slopes, unleashing a terrifying debris flow that moved at about 12 feet per second and killed sixteen people, most of them at a church camp in the San Bernardino Mountains. Hundreds of thousands of cubic meters of rock and soil were carried out of the mountains by the storm, enough to cover a football field hundreds of feet deep and to cost millions of dollars to remove.

Nor does the rainfall have to be as heavy as that which fell on Devore. A study of twenty-five post-wildfire debris flows in Colorado by a team of scientists reporting in the journal *Geomorphology* found that some of the flows were triggered by less than ten minutes of rainfall. Furthermore, 80 percent were unleashed by storms that lasted less than three hours, with

Following the 2013 Beaver Creek Fire in south-central Idaho, numerous debris flows were triggered by heavy rainfall, including this one in Badger Gulch. Increasingly, rainfall is closely monitored in the weeks and months following a wildfire, in an effort to anticipate—and therefore moderate—severely damaging debris flows.

most of the rain from those storms falling in under an hour. And the dangers posed by debris flows may persist for years following the initial wildfire event. It remains to be seen what effect climate change will have on debris flows; many computer models predict warming temperatures will cause more rain to fall on many parts of the globe. And in some places—those with steep terrain and prone to wildfire—even a little extra rain could have devastating effects.

At the same time, severe fires like we see burning today can result in large areas of so-called hydrophobic soil. When the organic material is burned from a forest floor in very hot fires, the soil can end up impervious to water. In part this happens because a great many ground plants have wax-like substances on them to minimize water loss; if a fire is hot enough and burns long enough, those substances end up being vaporized and then later, as things cool, congealing on the ground. Water-repellent soil keeps much-needed moisture from getting to groundcover plants and young trees that are trying to grow in the months and years following a forest fire, greatly slowing recovery.

It was partly as a result of hydrophobic soils that in 2002 Colorado's Hayman Fire loosed tons of ash and sediment flow into Cheesman Reservoir, which supplies Denver with 15 percent of its water. In that same year, massive trout die-offs in the Animas River in southwestern Colorado resulted from landslides and sediment releases after the Missionary Ridge Fire scorched more than 71,000 acres. In the aftermath of the Las Conchas Fire in New Mexico, instead of soaking into the soil the rain simply ran off, causing massive flooding and landslides, devastating local human communities for years, including 16,000 acres of the beautiful Santa Clara Pueblo.

Nor did the Las Conchas burn spare Bandelier National Monument. Anticipating the devastating effect of post-fire floods, park managers surrounded the visitor center with some fourteen thousand sandbags, as well as Jersey barriers, the low concrete walls you often see on highways to direct traffic away from construction activity. Water-repellent plastic was placed over the outside walls of historic buildings in the bottom of Frijoles Canyon, while bridges over Frijoles Creek were removed to keep them from becoming damming devices during strong floods. And strong floods are exactly what the national monument got, beginning with heavy rains in the Jemez Mountains on August 21, 2011. While the visitor center was mostly spared, popular trails—including the section of the beautiful Falls Trail that led from the Upper Falls to the Rio Grande River—were washed away.

And flood events following wildfires have other, even more critical impacts. Following the Las Conchas Fire, an Interagency Flood Risk Assessment Team concluded that the contaminants in sediments and ash transported by spring floodwaters could pose a health risk. People eating vegetables grown

DEERE
333D

in soil that had been treated with ash from the fire were thought to be at higher risk of certain cancers, the danger coming from ingesting radioactive strontium-90 and thallium. At the same time, researchers determined that merely taking regular walks through the ash-laden sediments of floodplain areas could result in overexposure to radioactive cesium 90. And finally, following the fire, fish in Lake Cochiti were showing elevated levels of mercury, probably from ash coming out of the forests during flood activity and settling on the bottom of the reservoir.

IN RECENT YEARS SCIENTISTS HAVE STARTED USING SATELLITE technology to help identify severely burned areas at high risk of flooding and landslides. This information is then fed to special Forest Service Burned Area Emergency Response (BAER) teams, composed of geologists, ecologists, wildlife biologists, engineers, soil scientists, botanists, and foresters. These women and men are charged with deciding whether seasonal rains falling on the burns might lead to significant danger, and if so, what to do about it. And more often than not, there's a lot riding on their expertise.

On deeply damaged soils, BAER teams may launch emergency seeding and mulching operations to prevent erosion. This work often involves laying down seed and then covering it with straw, crimping the straw into the soil, and then laying plastic netting over it to keep it from blowing or washing away. Over the years we've learned that such stabilization efforts need to focus on the use of native plants. While nonnative grasses and forbs may

TOP Land managers test for hydrophobicity following a severe fire in New Mexico's Santa Fe National Forest. When especially hot fires reduce the ability of soils to absorb moisture, it can lead to slowed plant growth, as well as greatly increase the chances of flooding.

BOTTOM Hay bales are loosened as part of an operation to drop 1,800 tons of straw by helicopter on remote lands burned by the Hewlett Gulch Fire of 2012. The effort was in large part to reduce soil and ash runoff into Seaman Reservoir, which provides drinking water to the city of Greeley, Colorado.

A soil scientist on a Burned Area Emergency Response team at the Rim Fire in California determines the depth to which the soil has been burned.

do better initially and thus stabilize the slopes sooner, they can also take over a landscape, squeezing out the plants that have been growing there for thousands of years.

In other areas, BAER teams fell and place logs on the contours of slopes to hold back moving soil and debris. When trees aren't readily available, such as on the chaparral slopes of southern California, workers may instead place 10-inch-diameter, 15-to-20-foot-long tubes of nylon mesh filled with straw (referred to as wattles), or even blocks of straw tied with twine and pinned to the ground in rows with wooden stakes. The basic idea is to interrupt the path of water flow and thereby slow it, which means installing such barriers in staggered tiers so that the center of one is directly downslope from the gap in the barriers above it.

In 2000 when a prescribed burn in northern New Mexico went terribly

The massive Trinity Ridge Fire in Idaho's Boise National Forest burned 146,000 acres in 2012. Rehabilitation programs were launched immediately to protect watersheds vulnerable to erosion.

out of control, a portion of the resulting Cerro Grande Fire moved into a watershed above the Los Alamos National Laboratory where radioactive waste from the 1940s had been buried during the development of the atomic bomb. A BAER team was dispatched to the area—charged, as usual, with making a thorough assessment of the risk within seven days of a fire being contained. Among other things, the team ordered the quick construction of a small dam in Pajarito Canyon to keep radioactive waste from washing off the site.

MUCH OF WHAT WE KNOW ABOUT LANDSCAPES HEALING AFTER A major burn comes from intensive research following the great Yellowstone fires of 1988, which burned nearly 8,000 acres in and around the park. As

Debris catchers were installed in many drainages following severe fires near Los Alamos, New Mexico, in an attempt to keep rocks and trees from causing damage as they were hurled downstream by heavy seasonal rains.

pundits across the country were busy decrying that America's first national park was forever lost, sure to be a sad skeleton of its former self for generations to come, scientists were quietly going about documenting the process of recovery. Yellowstone's ubiquitous lodgepole pine was showing signs of having reseeded itself well before a single year had passed, along with trees and shrubs with underground stems that sprout quickly, and also plants like fireweed with seeds that wait in the soil until fire passes over.

In fact, one Yellowstone research project, conducted the year following the big fires, found lodgepole pine seeds—nourished by nutrients leached from the ashes—sprouting at the extraordinary rate of three hundred thousand per acre, nearly all of them growing within a short distance of their parent tree. Thus what was lodgepole forest before 1988 was fast on its way to being lodgepole forest again. In truth not only were transitional

Firefighter Gillian Bowser surveys the damage near Lava Creek in Yellowstone at the end of a long day during the fires of 1988.

plants—fireweed, currant, and raspberry—abundant within just a couple of years of the burn, but also the vast majority of greater Yellowstone was on its way to rebuilding the same mix of vegetative communities that were present before the burn.

By spring 1989, woven in and around the seedlings were sprigs of blue-eyed Mary, prairie smoke, and sticky geranium, not to mention thousands of acres of nutrient-rich grasses, prized by bison and elk. Bicknell's geraniums, extremely hard to find the year before, rose from the ashes in profusion. By the time two years had passed, an abundance of flowering was going on, and shrubs were already proliferating, having resprouted from underground root stems. Indeed, while most biologists had predicted that such regrowth would depend on seeds dispersing from unburned areas of the forest, much of the new life actually came from individual plants in the burned areas that had survived the flames.

Once sediment washed out of the streams, increased light in the treeless landscape led to more algae, which in turn led to more insects. And all those

Young lodgepole pines resprout quickly and were well on their way to reclaiming the Yellowstone landscape just a few years after the massive fires of 1988.

bugs meant more food for trout. Three-toed woodpeckers, rare before the burns, were commonplace in the charred, insect-infested trees; tree swallows and mountain bluebirds flourished. True, certain nesting birds, as well as species favoring old-growth forests, such as pine martens, were displaced. Yet by all indications, only two species suffered significant population declines: capshell snails and moose, the latter having lost the mature spruce-fir forest so important to its winter range. As the years went on, researchers found no shortage of evidence that the fires, devastating as they may have seemed at the time, did not lead to any sort of ecological catastrophe.

BUT AGAIN, A PRESSING QUESTION IS WHETHER WE CAN COUNT ON this level of healing in the future. In some places, including California and much of the Southwest, such full-scale recovery may be part an outdated paradigm. As burns get bigger and hotter, and at the same time come with increasing frequency, some of the regenerative dynamics we saw in the wake of those 1988 Yellowstone burns may start to unravel—in part because such conditions make it difficult for the seed source to survive. In a major wildfire that took place in 2001 near Denver, 20,000 acres of ponderosa pine forest burned so hot that some fourteen years later, the trees still hadn't regenerated.

What natural communities will arise in the wake of giant, climate change–driven burns remains to be seen. In some places, what were once forests may start turning into perpetual grass and shrub lands. In fact, we can get a strong sense of such a change in New Mexico's Jemez Mountains in the wake of the 2011 Las Conchas Fire. In its first thirteen hours, that blaze consumed a phenomenal 44,000 acres—in other words, just under an acre every second. Swelling into what until that time was the largest fire in the state's history (a record broken the following year), by the time the Las Conchas Fire was contained in August, it had burned more than 156,000 acres.

Fueled by thousands of acres of insect-killed, tinder-dry pinyon pine, the Las Conchas inferno left few remaining seed sources, or for that matter, even nutrients. Five years after the burn, much of the landscape remained stark, with little vegetation to be found save cheatgrass and scrub oaks, which has led some researchers to speculate that what was formerly forest may become permanent grassland, interspersed with heat- and drought-tolerant shrubs. If climate models are correct—that conditions throughout the rest of the century will continue drier than they've been for a thousand years—what were once timberlands here, and in many other places in the West, will yield to a far more Mediterranean-looking landscape.

When shrubs take over the landscape, they may be around for a very long time. In part this is because plants like scrub oak are well suited to drought; but also, the shade they cast can make it difficult for native pines to germinate. It's not that these sorts of transitions haven't happened before.

But the types of disturbances being caused by megafires, combined with drought, have sped things up. Conditions today no longer reflect earlier natural processes, when certain plants died out over time—usually many centuries—to be slowly but surely replaced by others more suited to the emerging climate.

Paleoclimate records (records of past climates reconstructed from imprints such as tree rings and ice cores) suggest that dry periods have occurred in the Southwest several times during the past thousand years. However, rather than being the result of human activity, they were part of the climate's natural variability. But again, these changes tended to happen over long periods of time. The changes that have been recorded since 1965—like depleted soil moisture levels—are unprecedented. Flowing portions of dryland streams are in some places expected to decline by as much as 20 percent by 2050. This means more and longer stretches of dry channel, with potentially dire consequences for spawning native fish, as well as for a host of other wildlife that now use seasonal refuges.

What can we do? How can we both lessen the risks associated with the most devastating wildfires and manage the post-fire landscape to promote the greatest healing?

RISK REDUCTION

The art and science of prevention and treatment

WHILE THE ECOLOGICAL CONSEQUENCES OF WILDFIRE ARE ENORmous, so too are the financial costs to society. Of course there's the expense of fighting a major wildfire: the salaries of fire crews and other emergency response teams, as well as enormous aircraft costs, including aviation fuel and fire retardant. But that's only the beginning. Here's a breakdown of the largest burn so far in Colorado's history, the 138,000-acre Hayman Fire, which started about a hundred miles southwest of Denver in early summer 2002 and destroyed 133 homes:

Loans and grants from the Federal Emergency Management Agency: $4.9 million
Damage to power lines: $880,000
Private property loss: $38.7 million
Loss in National Forest recreation dollars to local businesses: $382,000

Value of timber lost: $34 million
Water storage capacity losses: $37 million

The total cost, then, of just this one fire was nearly $116 million.

And containment costs are going up. While as a nation we were spending roughly $600 million annually in federal dollars to fight wildfires back in 1995, twenty years later that cost had soared a whopping 500 percent, to about $3 billion. State government expenditures for wildfire suppression have soared, too, doubling between 1995 and 2015 to more than $1.5 billion. In 2015 in the state of Washington alone, the cost of dealing with some fifteen hundred wildfires rang up at more than $250 million.

At the same time, extraordinary secondary costs often result from fire, including lost revenues for businesses, damage to infrastructure, and revegetation expenses. Following Colorado's Buffalo Creek Fire in 1996, floods from an intense thunderstorm caused more than $2 million in damage to private property, plus another $20 million in damage to the Denver water supply system. In the first couple years following Colorado's Waldo Canyon Fire in 2012, the Forest Service spent more than $75 million just to help stabilize and recover some of the most severe burn areas.

Despite the overwhelming fact of hundreds of millions of dollars being spent every year on fire suppression and recovery, research suggests that shuttling some of those funds toward prevention—either to help homeowners reduce fuel around their homes or for land managers to thin or do controlled burns in forests with heavy fuel loads—would by far be a more efficient use of tax dollars. Notably, the Forest Service spends about five hundred times more on putting out fires in the wildland-urban interface, the WUI, than it does on preventing them in the first place.

This brings us to the conundrum that because we pay for firefighting from the same operating funds that Congress appropriates for research at a wide variety of agencies—including the Environmental Protection Agency,

Prolonged drought, along with a proliferation of invasive grasses throughout much of the West, has left many landscapes more fire prone than ever before. This prescribed burn in Zion National Park was done to reduce the risk of wildfire damage to the park's campground facilities.

the Forest Service, and the Smithsonian—we're rapidly losing what the Nature Conservancy's Christopher Topik calls science capacity. Since 2001, largely due to budget cuts in the wake of fighting fires, the research-and-development wing of the Forest Service has lost a full third of its staff. "There is much at stake with this loss of land management and science capacity," says Topik. "Forestlands provide half of our nation's water and sequester about 13 percent of total U.S. fossil fuel carbon emissions, but projections suggest that forests will become net carbon emitters later this century if steps are not taken to make them more resilient." Those steps will need to be driven by sound science—science that can lead to better fire behavior models as well as to more effective treatment strategies.

A hopeful development along these lines is the National Cohesive Wildland Fire Management Strategy. Developed over the course of four years with the input of every aspect of government, local to federal, the strategy focuses on three critical goals: first, to improve the ability of firefighters to respond to such emergencies; second, to restore fire-adapted landscapes (in other words, undoing some of the damage caused by eighty years of aggressive fire suppression); and finally, to create fire-adapted communities (towns and subdivisions in the WUI that will be far less prone to catastrophe from wildfire). These latter two goals alone—restoring fire-adapted landscapes and creating fire-adapted communities—would in the long run save billions of dollars that will otherwise end up going into suppression.

WHAT, THEN, CAN BE DONE TO ADDRESS THE FIRST OF THESE THREE goals—improving firefighters' ability to respond to a rapidly growing slate of wildfire emergencies? Of course, part of the answer is to make sure that we have enough firefighters and that those men and women have the best training, equipment, and technology at their fingertips. But there are other, less obvious steps we can take as well. For one, time and again it's been shown that among the most critical aspects of safe and effective wildfire response—and an area where mistakes continue to occur—is frequent, effective communication between different members of the various responder organizations.

In one of the most comprehensive research efforts to date, in 2010 Branda Nowell and Toddi Steelman intensively studied communication exchanges during three significant wildfire events in the WUI: the Bull Fire in California, the Tecolote Fire in New Mexico, and the Schultz Fire in Arizona. The key finding of their investigations is that communication was most effective among wildfire responders who knew one another. While emergency response protocols have been developed in military-style, hierarchical fashion, with a clear chain-of-command structure, how well people actually share information from an avalanche of rapidly changing, even chaotic conditions, depends on how familiar they are with different members of the larger team.

As it turns out, effective communication was most problematic between responders who shared similar roles in different agencies but didn't know one another. This breakdown could in some cases have to do with trust—being confident that your cohort is going to perform as well as you'd like. But more often it seems to have to do with the fact that it's awfully easy to assume someone in your same job in another agency sees the world, and solves problems, pretty much like you do. The solution suggested by the authors seems simple, but in fact it's quite profound; fostering better communication among emergency wildfire responders, they concluded, may be no more complicated than providing opportunities for firefighters to spend time together, both professionally and socially, before incidents occur.

Other communications issues are being addressed as well. For starters, individual fire districts are finally adopting so-called common communication plans, which assure that firefighters from different agencies are all plugged into common radio frequencies so that everyone has full access to what's going on at any given time. Beyond that, in states like Montana and Arizona, agencies like the Montana Department of Natural Resources and Conservation (DNRC) and the Arizona Department of Forestry are making efforts to improve communications in a way that allows fast help for local fire departments. "Fires that we don't get to quickly," says incident commander Jon Trapp, "can blow up into major burns." Keeping that in mind, the Montana DNRC, for example, now monitors all the fire frequencies in the state. If a call is made to 911 reporting that someone has spotted smoke—an alert that's usually passed along to the local rural fire department—DNRC calls the local team to see if they need anything, be it bulldozers or engines or hand crews or even strategic planners.

"For us to gain even a 1-percent increase in efficiency," says Trapp, "could mean saving tens of thousands of acres from burning, as well as millions of dollars." Trapp says the idea is to get enough resources to a new fire as fast as possible. "The cost of front-loading a wildfire with equipment and personnel in the first twenty-four hours," he explains, "is minuscule compared to what will be spent if it takes three weeks to suppress it."

WHAT ABOUT THE SECOND GOAL—THAT OF RESTORING FIRE-adapted landscapes? It's here that range, forestry, and ecological scientists around the world are working at a feverish pace. As we saw earlier, a staggering number of acres in the western forests have excess fuels thanks to eighty years of fire suppression. Not only is there downed timber, which in arid climates can take decades to decompose, but there are also crowded forests with high tree density as well as an abundance of so-called ladder fuels that allow what would normally be ground fires to rise into the canopy. Finally, more open forests and rangelands are in many places being overwhelmed by invasive species such as cheatgrass, which alters normal fire patterns even more.

Part of the effort to recreate fire-adapted landscapes, and thus reduce the severity of wildfire, means putting money into so-called forest treatments. One such treatment consists of thinning the woods, which essentially reduces the intensity of fires by reducing the number of trees—eliminating young, tightly spaced timber, which tends to be especially flammable. At the same time, thinning has the advantage of reducing competition for dwindling water resources; that, in turn, keeps the remaining trees healthier and thus more resistant to insects and disease, which in turn keeps them from turning into standing tinderboxes. What's more, by decreasing the size of the canopy, thinning ensures that more of the rain that falls will reach and soak into the ground.

Ladder fuels, like this tree going up in the Cascade Complex Fire in central Idaho in 2007, provide a way for ground fires to rise into the canopy. To reduce the risk of canopy or crown fires, so-called forest treatments—either prescribed burns or selective logging—are being proposed in many parts of the West.

But research suggests that to have the desired effects, thinning has to be done on a fairly large scale, removing a significant number of trees across a large number of acres. At an estimated cost of $28 billion to treat all the places that currently need it, such efforts would break the bank. Instead, foresters will have to limit such efforts to high risk areas, where fuels meet structures.

It's important to note that thinning isn't universally thought to be a good idea. Some researchers contend that because canopy reduction exposes the ground to sunlight, it not only increases evaporation but also creates perfect growing conditions for grasses and forbs that in turn end up competing with the trees for water. Reducing the tree canopy by thinning can also cause the temperature in the forest to increase, the humidity level to decrease, and fuels on the ground to desiccate more quickly—all of which could lead to more severe burns. In addition, the roads needed to remove logs from a forest for commercial use can be both an entry point for weeds and a path for sediment runoff into local streams. Yet even with all these cautions, at this point the majority of scientists seem to agree that thinning—especially when followed up with prescribed burning to clean up ground plants and debris—can in some places be a useful tool for dealing proactively with wildfire.

This brings us to the other treatment method, prescribed burns, where downed logs, small trees, and other flammable fuels are cleared away by means of controlled fires. Prescribed burns are arguably among the faster ways of getting a forest back to a healthier, more natural state. But like thinning, to have much of an effect prescribed burning has to be done on a large scale. When discussing the Yellowstone fires of 1988, one researcher suggested that reducing the severity of those fires to any considerable degree would have required prescription burning of about 50,000 acres a year for nearly a decade. So far the public has been anything but enthused about that level of burning—in part because of the health consequences of so much

After thinning operations, wildfires moving through this area are likely to be more modest in size.

smoke being released into the air. Indeed, in some states, Washington being perhaps the best example, public land commissions are reluctant to allow prescribed fires, even on national forest lands, citing vigorous public complaints about the resulting smoke.

With that in mind, a cutting-edge modeling tool being developed at the Missoula Fire Sciences Laboratory will likely play a big role in helping managers decide the right time for prescribed burns. "If you're planning a prescribed burn on the north side of Mount Hood," explains scientist Bret Butler, "you have to be really concerned about the impact on communities in the Columbia Gorge." Butler says that currently the only prediction tools for assessing smoke transport and dispersion are based on Weather Service models, which offer only one data point for about every 4 square miles of land. But the model being developed by Butler—part of the tool he and his

colleagues developed to help predict fire spread from wind—will operate on a much finer scale and will, they hope, "increase our ability to keep smoke from prescribed burns out of population centers." Prescribed burns may in fact be an area of fire management where complex modeling is at its most useful—allowing managers to better understand the exact conditions under which a controlled fire can be set in a given forest.

Related to this is the practice of managed burning, which involves letting a natural blaze started by lightning just go ahead and burn—again, with the goal of reducing fuel loads in the area. Of course, the decision to let a lightning-caused fire burn is hardly taken lightly. For starters, is the weather right—high enough humidity, low enough temperatures, and calm enough winds? Is the location right—not only far enough away from homes and other structures but also in terrain where fuel loads are either light enough or moist enough to not lead to major conflagrations? And what if something goes wrong? Are there features in the area—roads, creeks, rock ledges, meadows—that can check the burn if it starts to run? Once again, besides drawing on their own considerable experience, at this point wildfire teams have painstakingly crafted incident response guides to help them make such decisions, not to mention having at their disposal a bevy of computer-based programs offering such things as topo maps, localized weather, and fire spread prediction models.

The financial cost of prescription burning ranges from about $12 an acre for a 100-acre slice of relatively flat Douglas-fir forest (well away from any homes) to fifteen times that per-acre cost for 200 acres of mixed conifers near developments. But such burns aren't without risk. As is so often the

TOP **Prescribed burning in a wheatgrass-sagebrush-juniper community in eastern Oregon reduces the chance of a hotter, more catastrophic wildfire in years to come.**

BOTTOM **To reduce the risk of prescribed burns accidently spreading beyond the intended fire zone, land management agencies often burn piles of brush during winter months, when fire danger is at its lowest.**

case when putting science to good use on the ground, there's a lot more to this than just lighting a fire.

Though it certainly doesn't happen often, sometimes prescription fires get out of hand, including an ill-fated prescribed fire in 2000 that was started in New Mexico's Bandelier National Monument and got away due to high winds and drought conditions. The Cerro Grande Fire, incredibly, ran up a tab of nearly a billion dollars in firefighting costs and property damage. This was the fire mentioned in the previous chapter that moved into the watershed above the Los Alamos National Laboratory. Eighteen thousand residents of the Los Alamos region had to be evacuated, and several hundred families lost their homes.

Prescribed burns have also sometimes had the unintended consequence of opening the landscape to invasive species. For example, in the ponderosa pine forests of California's Kings Canyon National Park, fire managers began a series of prescribed burns meant to mimic the natural historic pattern of frequent, low-intensity fires. But they made a mistake by lighting the burns too frequently. As a result the canopy stayed open, which allowed sunlight to hit the forest floor; at the same time, not enough essential nutrient litter accumulated in between burns. The result was to create perfect conditions for a massive invasion of cheatgrass.

In a number of other cases, managers planted pine seedlings following prescribed burns while at the same time using herbicides to kill native shrubs they thought would compete with the young trees. But absent native pioneer species—this due to the herbicide—what came in instead were thick mats of highly flammable invasive grasses. These later ignited and burned so hot in each case that the fire destroyed the entire seedling plantation. Finally, it's lately become apparent that doing prescribed burns out of season, say in late winter or early spring, with the intention of reducing the chance of unintended wildfire spread, can open up the land to massive alien plant invasions. The reason for this seems to be that such burns expose native seeds to moist heat (as opposed to dry heat), which for many is deadly.

Yet for all these cautions, both prescribed and managed burning are critical tools for decreasing the risk of catastrophic wildfire in the decades to come. During the often terrifying fire season of 2015, for example, in the Okanogan-Wenatchee National Forest of Washington, the North Star Fire—responsible for burning hundreds of square miles of national forest—was blasting right toward homes in the Aeneas Valley. In the end what kept those homes from being lost was the fact that Forest Service crews had earlier in the spring and fall seasons burned the intervening 850 acres of ponderosa pine forest to eliminate undergrowth. Making a stand during the North Star Fire in this previously burned area, firefighters were able to stop the fire's advance. Likewise during California's Rim Fire in 2013, the fact that the Forest Service had done several prescribed burning projects near the Whiskeytown National Recreation Area slowed the fire's advance enough in that area to allow firefighters to get out ahead of the burn—and in the process, save a number of homes from going up in smoke.

Mastering the art and science of prescribed burns, then, means learning as we go, and now and then making mistakes along the way. Such wrong turns, rather than being a sign of carelessness, are a humble reminder that there's still much for us to learn about ecological processes. The more committed we are to funding prescribed burn projects—and also, the more we support gathering data from such efforts—the more effective our efforts will turn out to be. But it's a big job ahead of us; and at this point the amount of land that needs prescribed burning or thinning continues to grow with every passing year. In California alone, about 15 million acres were in need of treatment in 2016.

THE GOOD NEWS ABOUT THE THIRD GOAL OF THE NATIONAL COHESIVE Wildland Fire Management Strategy—creating fire-adapted communities—is that wildfire catastrophe in the wildland-urban interface can usually be prevented. Most often the homes that burn down in wildfires do so from things like homeowners having landscaped with mulch up to the edges of the house. Or having failed to clean leaves and pine needles from gutters. Or having an unscreened attic vent, or an open trash or debris can sitting against an outside wall. Even unsealed pet doors can lead to catastrophic ignition.

Groups like the National Fire Protection Association are working overtime to educate builders and homeowners in the WUI about how to reduce the risk of homes catching on fire. They preach the gospel of fire-resistant roofs. Of using stone siding along the bottom of a house instead of wood, or opting for nonwood paving stones in lieu of wooden decks. They're encouraging homeowners to place a simple metal cap across the gap between the underside of the roof and the horizontal trim board, or fascia, thus preventing embers from being drawn into the roof structure. They talk of covering crawl space and attic vents with screens. Of removing dried grasses and flammable shrubs like bitterbrush from around the perimeter. Of pruning low-hanging limbs from evergreens, as otherwise the limbs provide a ladder for fires to climb from the ground up into the tops of the trees.

In Colorado's devastating 2013 Black Forest Fire, near Colorado Springs, nearly five hundred homes were lost. In one area of that community alone, sixty-one out of sixty-seven homes—many built in the late 1920s—were destroyed, in large part because the woods had been allowed to grow more or less unchecked. And yet in nearby Cathedral Pines, where fire-oriented master planning and prevention had long been a top priority, less than a handful of structures were lost. Because homeowners had taken precautions—placing rocks around their houses, keeping mulch away from the edges of buildings, making sure yards were free of trees and other flammable vegetation—the advancing flames were slowed. That, along with the

TOP TEN WAYS TO PROTECT YOUR PROPERTY FROM WILDFIRE

These ten simple yet highly effective steps (in no particular order) can protect homes from wildfire.

1. Maintain defensible space in the critical area between 0 and 5 feet from your home by using noncombustible materials such as gravel, brick, or concrete.

2. Maintain defensible space in the area 5 to 30 feet from your home by removing dead vegetation, removing shrubs under trees, pruning branches that overhang the roof, and thinning trees. Trailers/RVs and storage sheds in this area should be moved or have defensible space around them.

3. Reduce siding risks by using noncombustible siding or maintaining a 6-inch ground-to-siding clearance.

4. Use a roof covering fire-rated Class A, which offers the best protection for homes.

5. Regularly clean debris from the roof, since debris can be ignited by wind-blown embers.

6. Regularly remove debris from gutters for the same reason. Gutter covers, if used, should be noncombustible.

7. Use noncombustible materials for fences and gates to reduce the risk of burning fencing catching your house on fire.

8. Cover vents with ⅛-inch mesh and box-in open eaves to keep burning embers out of your attic.

9. Protect your windows by using multi-pane tempered glass, and close the windows when a wildfire threatens.

10. If you have a deck, use boards that comply with California requirements for new construction in wildfire-prone areas. Also remove combustibles from under the deck and maintain defensible space.

For those living in the wildland-urban interface, it's never been more critical to minimize ignition potential in homes. Smoke plumes from the 2014 Mills Canyon Fire in the Wenatchee National Forest in Washington were visible from residences that stood their best chance with fire-resistant roofs and defensible space around their perimeters.

fact that a careful thinning of vegetation had kept fire on the ground instead of allowing it to reach the tops of trees, allowed firefighters the time they needed to save the homes.

And there's more that can be done. Local and federal agencies would do well to invest money in helping owners achieve basic alterations on their homes or to provide things like pickup service for annual tree and brush trimmings. Though every summer municipal firefighters do incredibly impressive work (again, wildland firefighters themselves are rarely charged

In just the first two days of Colorado's Black Forest Fire, more than 360 homes were destroyed and more than thirty-eight thousand people were evacuated. The cost to fight the fire exceeded $9 million, with an additional $90 million in property damage and loss.

with structure protection), they'll never be able to stop every major burn from destroying homes and businesses. With that fact staring us in the face like never before, it's time to start living wisely with the inevitability of fire.

Unfortunately, we've got a long way to go. In the devastating Yarnell Hill Fire of 2013, for example, only 13 of the more than 120 structures in the path of the burn had been properly protected by their owners. As of 2016, of some seventy thousand communities that have been designated as being at high risk for wildfire damage by the National Fire Protection Association, only about nine hundred, or slightly more than 3 percent, have taken steps to reduce the danger.

As more and more people stream into the WUI, as fires increase in both number and intensity in the face of climate change, local governments are going to need to step up and enact land use planning precautions. Given that

The *Los Angeles Times* has called Malibu, California, "a virtual laboratory for wildfires," with homes on hillsides surrounded by dry chaparral and north-south canyons that funnel hot, dry Santa Ana winds toward the Pacific.

among the best predictors of property loss in the face of wildfire is whether homes are less than thirty feet apart, it's unconscionable that new housing developments in the WUI aren't being constructed according to National Fire Protection Association standards. Besides requiring new developments to meet these standards, local governments can prohibit flammable building materials like wood shingles, discourage overplanting in the so-called structure ignition zone (a space extending out 200 feet from the perimeter of a home), and develop community wildfire protection plans.

When deciding on style, density, and location of new subdivisions, developers need to take into consideration things like water supply, how effectively people can get into and out of the subdivision, and the capabilities of the local fire department. Many rural housing areas in the West have been developed one home at a time, with lots later being split into smaller parcels. Over the years this sort of piecemeal development has led to densities similar to that of suburban subdivisions but without any of the protections planned subdivisions commonly offer. For example, water supplies may be limited—and often woefully inadequate to fight wildfire—and the access into the area may be severely constrained by narrow, winding roads, sometimes with severely weight-limited bridges.

With wildfire here to stay, land use regulations throughout the West should be updated to include fire-relevant building codes. And in extremely fire-prone areas, development should be seriously restricted. As we've learned from other natural disasters like floods and hurricanes, making questionable landscapes safe for humans in the short term often has the unintended consequence of making devastating losses more likely sometime down the road.

THE REMARKABLE EFFECTS OF HUMAN-CAUSED CLIMATE CHANGE, along with overzealous suppression of wildfire, has left biologists, ecologists, and foresters scrambling to figure out what to do next. Is it wise to be proactive following big burns, even to the point of helping establish more drought-tolerant native plants than were there originally, thereby minimizing the risk of massive invasions by exotic, nutrition-poor weeds? Or could there be unintended consequences from kick-starting such transitions?

Some of the most prickly debates on that question are arising around the management of federally protected wilderness. As some biologists predicted way back at the start of the twentieth century, America's wildlands have given us an invaluable set of baselines for showing us what healthy natural systems look like. These baselines have already provided scientists with the knowledge they need to restore damaged landscapes—from reestablishing salmon fisheries in the Northwest to reclaiming toxic mine sites in California and the Rockies. At the same time, wilderness areas are teaching us about the fundamental needs of natural systems in the face of increasing drought, invasive species, wildfire intensity, and other climate-change-related processes.

Taking full advantage of those lessons, though, may require two things. First, given that some animal species won't survive in the face of the rapidly shifting habitats that climate change induces, we'll need to expand the size of some of our current wildland preserves. Connecting landscapes with north-south corridors to create better migration opportunities is among the most widely endorsed ecosystem management strategies in the United States today. Yet even if that happens—and the political climate is far from friendly to such notions—many species simply won't be able to migrate quickly enough from their current habitats to more suitable ground.

Unfortunately, unless wilderness managers assist in those migrations, moving or transplanting threatened species to more appropriate habitats, those species are going to become regionally extinct. And this second action item is where things get really sticky. While the current Wilderness Act allows great flexibility—providing for all sorts of special actions as long as the original intentions of the act are honored—it can certainly be argued that relocating species to areas where historically they never occurred, is prohibited.

In the coming years the courts will hear a variety of challenges to the idea of wilderness managers being allowed to pursue such hands-on approaches to species preservation. On one hand, it's hard to fault those who object to what can be seen as meddling. As plant ecologist Frank Egler pointed out in the early 1980s, "Ecosystems are not only more complex than we think. They're more complex than we *can* think." Time and again we've found out the hard way that just when we thought we were being

Now, and for many years to come, climate change will create not just scientific but also philosophical challenges—both for wilderness resource managers and for those who find solace and comfort in these precious wild places.

helpful—as in the case of suppressing all wildfires—we were actually causing harm. With this in mind, even if the courts do grant increased power to wilderness managers to do things like assist species migration, it seems prudent to allow some preserves to continue under the more traditional hands-off policy. The challenge will be knowing what to do, and where.

When the Wilderness Act passed in 1964, it was to most people inconceivable to think that fifty years later every landscape on Earth would be deeply impacted by the effects of human-caused climate change. If wilderness is to continue to be characterized by healthy watersheds and vigorous biodiversity, it's likely that some of its provisions will have to be interpreted in more liberal fashion. The wilderness system was fashioned both from our notions of nonhuman life having the simple right to exist and from a desire to pass along these natural wonders to future generations. If we're to honor those admirable impulses, we'll have little choice but to use every thoughtful, well-considered tool at our disposal.

As for the millions of acres of forest and shrub lands that lie outside the wilderness, there's much left to think about there, as well. With warmer, drying temperatures entirely eliminating some plant species—which in turn results in loss of bird and mammal diversity—it's not yet clear what the best way is to intervene. Or is there even a best way? After all, preserving the current mix of species could be incredibly expensive. For example, the seedlings of some trees especially valuable to the ecosystem may have to be watered in order to survive; this is already happening in some places, including mountainous coastal areas of Spain, where following wildfires, seedlings are being kept watered with the use of special fog collectors. With the possible exceptions of trying to save iconic tree species, such as the sequoia, in most places we'll likely have to accept major shifts into new forest communities. But again, such transition will be shaped by fire.

CLEARLY, WE'VE STILL GOT A LOT TO FIGURE OUT. BUT THE INCENtives are high. Besides providing critical habitat for a staggering array of wildlife, forests also filter and store water, allowing snow and rain to percolate into underground aquifers. The estimated value of the water filtration and storage services provided by the earth's forests is more than $4 trillion a year; as a corollary, for every 10 percent reduction of forest land, the cost of treating drinking water grows by about 20 percent. Currently about two-thirds of the U.S. population depends on the water provided by forests. In fact, today some eighty-five thousand communities in the United States, or about 180 million people, rely directly on forests to capture and store the drinking water they use.

In addition, just the semi-arid woodlands of the United States like those in southern Colorado, New Mexico, and Arizona store carbon in amounts measured by the gigaton, which is the equivalent of a billion tons. Losing these forests would feed into yet another one of those self-reinforcing loops we face with climate change: lost trees resulting in less carbon being stored, which means more warming and drought, which then kills still more trees, thus feeding future fire.

FUTURE FIRE

SCIENTISTS AT NASA ARE ROUTINELY USING SATELLITES TO RECORD wildfire activity around the planet—observations that among other things allow them to note what regional climate conditions are like when such fires occur. Increasingly, that information is being combined with climate models, which are highly sophisticated mathematical predictions of what future climate conditions may be, to help us make intelligent estimates of what future fire activity is likely to be in various locations around the globe. And when it comes to western North America, the future is more than a little troubling: extreme events, like the extraordinarily dry conditions in 2012, 2013, and 2015, which in the past occurred on average about once a decade, are expected by the middle of the century to be happening three to five times as often. Related to this, the length of the fire season is likely to continue to expand.

Just as fire has long shaped the ecology of the American West and will for centuries to come, so will it increasingly shape the lives of human communities.

NASA satellites are being used to track wildfire activity around the globe.

The exact effects of climate-change-related megafires and prolonged drought on America's forests depend not only on the region we're talking about—the Pacific Northwest, or the northern and central Rockies, or the Southwest, or California—but also on the kinds of climate shifts occurring there. Not surprisingly, studies of fire tree rings from earlier epochs confirm the general relationship between warm, dry periods and wildfire. But the exact effects of that relationship depend on location. In the more widely spaced forests of the Southwest, for example, the most robust fire seasons were those with winter moisture, sometimes quite generous, followed by drought. In the northern Rockies, on the other hand, big fire seasons showed up in years of late spring and summer drought. The amount of moisture that occurred in winter didn't much matter.

A satellite image taken near Mount Wilson, California, detects the "radioactive power" of currently burning and previously burned segments of the landscape. Red-colored areas indicate fire happening as the image was being taken or within the previous twelve hours, while yellow areas burned six days prior.

ATMOSPHERIC PATTERNS HAVE A LOT TO DO WITH WHAT FIRE CONditions will be like in coming decades. To start with, wet and therefore less risky fire seasons come with El Niño ("the child"), or as climate scientists know it, El Niño Southern Oscillation, or ENSO for short. Among the most important global climate drivers on the planet, in any given year it can have a big impact on wildfire risk all across the United States. El Niño years favor robust vegetation growth, be it in grasses, shrubs, or trees; in the years that follow, these plants may dry out and become fuel for the big fires that tend to get purchase in the warmer, dryer conditions of La Niña years. One study for the National Academy of Sciences that matched La Niña years to fire-scar records found that large fires in the Southwest tended to occur during years when this particular climate event was in play.

Some scientists are wondering whether the El Niño cycle—which historically happens about every seven to ten years—might under climate change grow both more frequent and more intense. One thing's for certain: we're going to be hearing about El Niño and La Niña long into the future. So what, exactly, are El Niño and La Niña events? In simple terms, both are created by the movement of trade winds, which are winds that blow almost constantly in one direction.

In the more typical, non-ENSO event years, trade winds from the tropical oceans flow in a steady, predictable manner from east to west, pushing surface waters away from the southern Pacific coast of the Americas toward Indonesia. As the water travels, it heats up under the sun—and because of that heating, expands. In fact, this expansion is so substantial that the water resting at the western edge of the Pacific is more than a foot and a half higher than that along the coast of South America. These big pools of warming water, in turn, tend to create pools of warm air above them. That warming air reinforces the trade winds, which basically keep driving the cycle, in the process bringing lots and lots of rain to places like Singapore and Malaysia.

Sometimes, though, and for reasons not entirely understood, those westerly trade winds weaken. And when they do, the warm water in the Pacific begins traveling in the other direction, toward the east. Again, the warm water moving east warms the air above it—just like it did before, but now in the opposite direction—and that moving warm air keeps heat in the water below, creating a kind of feedback loop that leads to what's called a warm ENSO event, or El Niño. El Niño events typically begin in early fall and become strongest from midwinter through spring, not fading again until the following fall and winter.

This event, marked in large part by a weakening of the Pacific trade winds, ultimately leads to a rather complicated redistribution of heat along the equator. As it turns out, that redistribution of heat tends to cause all sorts of changes around the globe; in fact, outside of Antarctica, not a single

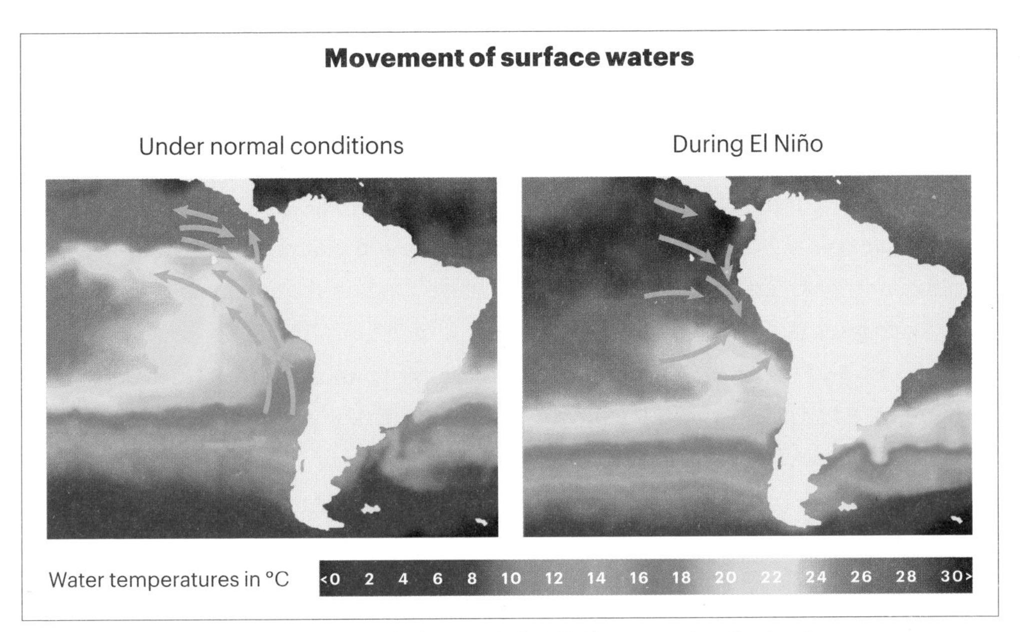

The movement of the trade winds changes during El Niño events, also affecting the movement and temperature of the waters.

continent escapes the influence of warm ENSO climate events. In the United States, for the most part El Niño causes those living in the northern tier of the country to experience warmer winters, and in many of those same places, drier as well. The Southwest, on the other hand, is usually wetter. The Pacific Northwest and California tend to see increases in winter precipitation.

La Niña, which is referred to as a cool ENSO event, involves the usual east-to-west trade wind flow patterns, but at such times the flows are much bigger—leading to an especially large temperature difference between the eastern edge of the Pacific at the Americas and the western edge of the ocean at Indonesia. When this happens, the northern states may be colder than normal during the winter, and the Southwest, often drier.

There's still a lot we don't know about the processes that allow big El

Niño events to evolve; the remote equatorial Pacific, where much of the action happens, is considered by many scientists to be a meteorological black hole. With that in mind, beginning in late January 2016—against one of the most extreme El Niño events ever recorded—climate scientists began a massive research effort involving specially equipped planes, weather balloons, and a large NOAA research staff, gathering data that will not only improve forecasts but also fine-tune existing climate models.

As important as ENSO events are, they're far from the only game on the planet. In the northern Pacific there also occurs something called the Pacific Decadal Oscillation (PDO), marked by twenty-to-thirty-year-long periods of either warmer than normal or cooler than normal water, which in turn affects temperature and precipitation levels from the Pacific Northwest to the Canadian prairies to the American Southwest. Researchers at Ontario's University of Guelph have discovered that in the Pacific Northwest, seven out of the ten years when there was a positive PDO index had bigger-than-normal fire seasons; similarly, eight of the ten smallest-fire years in that region happened when the PDO index was negative. And if all this isn't enough to make your head spin, there's also a pattern called the Atlantic Multidecadal Oscillation, which affects rainfall through central North America, and another called the Northern Hemisphere Annual Mode, which at this point seems to affect factors such as wind speed and the frequency of lightning.

INCREASINGLY, CLIMATE CHANGE MODELING IS USING THESE atmospheric patterns to help predict what fire conditions will be like in the coming decades. So far, nearly every such effort has produced data suggesting a substantial increase in wildfire activity in years to come—especially in the western United States and the Canadian boreal forest. That increase will be driven by hotter, drier conditions making forests more susceptible to the kinds of insect invasions that can kill trees (thus creating massive fuel loads for fires) and also making it more likely that wildfires will burn hot and fast.

The truth is there's no place on Earth not currently experiencing ecological transformations in the face of climate change. While the exact nature and degree of such changes are still quite difficult to forecast, our prediction skills are improving with every passing year. Many of the uncertainties that remain rest not on whether temperatures will increase (or in a few cases, decrease), but whether those changes will be accompanied by less snow and more rain, or less of both. The question of precipitation aside, current research suggests that a mean temperature increase of 3.6 degrees F (2 degrees C) by the middle of the century—and keep in mind that this is the target limit set in Paris during the 2015 climate talks—will increase the amount of area burned in the West by about 6 million acres annually, which is significantly bigger than the entire state of Connecticut.

From the Arctic to the Amazon, throughout much of Africa to Australia, from China to North America, there's an avalanche of evidence of the mounting consequences of climate change for the world's forests. The effects of climate change on western forests are more visible with every passing decade and include the decline of Colorado's famous aspen woods, the massive losses of pinyon pine and juniper that have already occurred in the Southwest, and the staggering breadth of dying evergreens up and down the Rocky Mountains. The number of trees in the West dying for no clear reason more than doubled between 1985 and 2015. What's more, a 2014 report by the Rocky Mountain Climate Organization and the Union of Concerned Scientists predicted that by 2060, aspen in a six-state area of the Intermountain West will decline by more than 60 percent. Perhaps more alarming still, landscapes suitable for sustaining the iconic conifers of the region—Douglas-fir, lodgepole pine, Engelmann spruce, and ponderosa pine—are expected to decline by fully half.

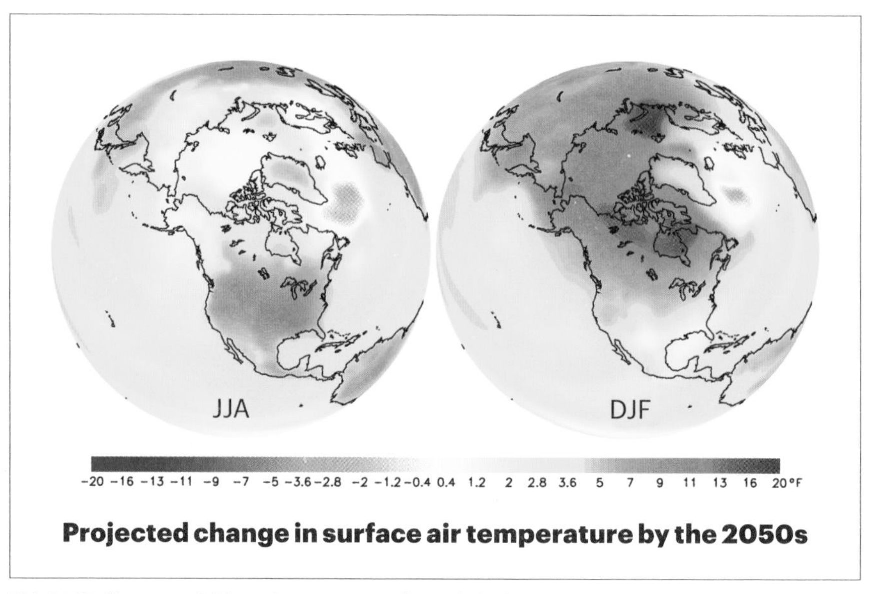

This 2007 climate model from the NOAA Geophysical Fluid Dynamics Laboratory shows projected surface air temperature changes between the late twentieth century (an average for the years 1971 to 2000) and the middle of the twenty-first century. The left panel shows expected increases for June, July, and August; the right panel, for December, January, and February. Temperatures in the western United States are projected to increase by at least 3.6 degrees F in the summer months.

But forests grow back, don't they? And besides, it's been shown that carbon dioxide is essentially food for trees and plants, allowing even mature forests to grow at a faster rate than without the extra CO_2. Isn't it possible that increases in carbon dioxide from anthropogenic climate change, along with a lengthening of the growing season, will create bigger, more robust forests than ever before? Not necessarily. Such so-called carbonization fertilization is a complicated business, especially when you try to apply it at the level of an entire ecosystem. While carbon dioxide does in some cases seem to increase plant production in food crops, when you reduce forest productivity—by drought, insects and disease, and wildfire—you end up with more ozone in the atmosphere, and ozone reduces crop production levels. In truth, it's a riddle we aren't even close to understanding.

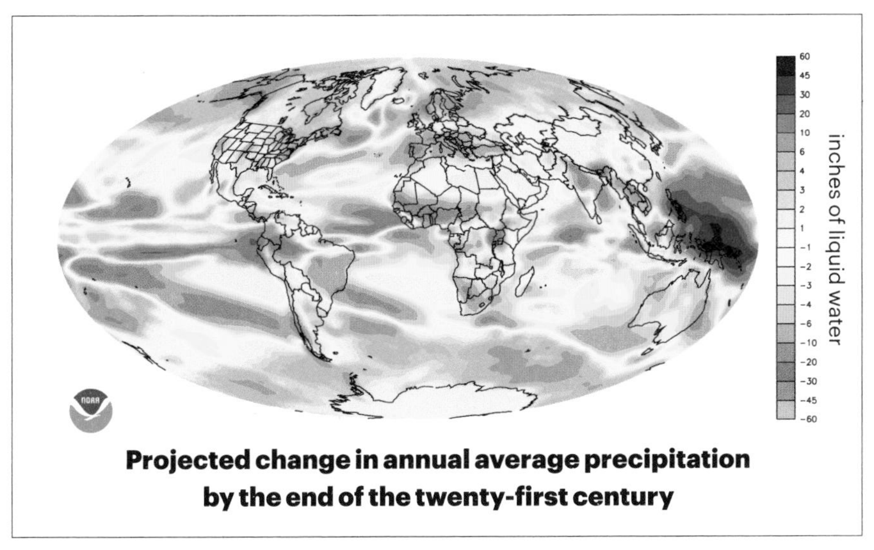

This 2012 climate model from the NOAA Geophysical Fluid Dynamics Laboratory shows the projected change in annual average precipitation by the end of the twenty-first century. It's expected that global warming will be accompanied by a reduction in precipitation in the American West.

What we do know is that for the long-term foreseeable future, drought and heat-related stress on trees will lead to large die-offs. With less rain, and at the same time with snowpacks melting earlier, there's often insufficient water available when trees most need it—at the height of summer. Short of killing trees, that stress can also leave forests more vulnerable to insects. As previously mentioned, in Arizona, New Mexico, and southern Colorado, dwindling tree health in the face of drought caused the death of tens of thousands of pinyon trees and is playing a major role in fostering the spread of the pine bark beetle in pine forests up and down the Rocky Mountains. Meanwhile, other insects like the hemlock woolly adelgid are expanding their range northward, able to occupy habitat that because of cold temperatures was previously off limits to them.

THE COMPUTER MODELS THAT PREDICT MORE FREQUENT BIG WILD-fire events by midcentury are basing those predictions on a scenario in which humans continue to release carbon dioxide into the atmosphere in amounts similar to what we're releasing now. Despite the enormous gains we've made in sustainable energy technologies, that scenario may not be far from the truth. Wildfires will in some places soon be contributing as much carbon over a couple months as the cars of those regions let loose in an entire year.

Normally, the planet's forests act as carbon sinks; they are currently responsible for absorbing and storing more than a quarter of all the carbon human beings are putting into the atmosphere—about 2.5 billion tons, which is an amount similar to what gets stored in the world's oceans. Unlike other kinds of plants, which store but then lose carbon in a quick process of decay, trees can store large amounts of carbon dioxide for decades, even centuries. But if forests are compromised to too great a degree, that storage capacity will diminish—and that, in turn, will further accelerate climate change in what climatologists call a positive feedback loop. Furthermore, dying and burning forests can become carbon sources, net contributors of carbon. Since roughly 1865, the loss of forests worldwide is thought to have contributed about 30 percent of the buildup of carbon dioxide in the atmosphere.

Currently, forests in the United States are able to offset from 10 to 20 percent of the nation's carbon emissions from fossil fuels. And for centuries the amount of stored carbon released during forest fires has generally been offset by the amount of new plant and tree growth happening elsewhere on the planet. But if fires get bigger, and if in some places forests are replaced by shrubs and grasslands, conceivably we could find ourselves in a situation where more carbon dioxide is being released than can be stored by the new growth. The carbon released by wildfires between 1996 and 2008 was already nearly 2.5 times more than that released between 1984 and 1995.

Carbon uptake by trees is also compromised by drought, which stresses a tree's physical mechanisms. Furthermore, this state of reduced efficiency

In California alone, so-called hot droughts, which are associated with human-caused climate change, have already killed millions of trees, especially in lower-elevation forests like this one.

doesn't end as soon as the drought ends. In fact, the effects can linger for two or three more years. Since the new millennium, the West has been more in severe drought than out of it. As a consequence, the amount of carbon taken up by our forests has dropped considerably; researchers estimate that between 2000 and 2004, carbon uptake declined by some 30 million metric tons.

RENOWNED ECOLOGICAL SCHOLAR C. S. HOLLING DEFINED THE idea of resilience in ecosystems—the ability to persist—as the amount of disturbance the system can endure without suffering big changes to its processes and structures. And when it comes to America's forests, this

Earth's forests are critical to human survival. In addition to creating essential fresh water reserves, they're also responsible for absorbing more than 25 percent of the carbon humans are putting into the atmosphere.

is the big challenge ecology must now put itself to—steering toward practical, on-the-ground management tools to help ease the consequences of our previous mistakes, while at the same time encouraging more creative, sustainable ways of living. Whether we're trying to reduce fuel loads in the big woods of the West, stopping destructive invasions of nonnative plants, maintaining wildlife diversity, or controlling wildfire-related erosion and landslides, we'll need to marry the most rigorous science we can muster with a kind of genuine humility and commitment that until now has too often been in short supply.

We continue to gain scientific information about wildfire. A lot of questions still remain, and a lot of essential science is yet to come. The Missoula Fire Sciences Lab, for example, is doing cutting-edge work on an amazing range of investigations. Scientists there are looking into, among other things, how low-intensity fires may induce long-term resistance to bark beetles in lodgepole pines, and they're refining our understanding of what it is that most puts homes at risk in the wildland-urban interface.

While technologies that allow safer, more efficient firefighting (such as modeling software to help the men and women on the fire line to predict fire behavior) will be critical in the years to come, science may prove most essential when it comes to early detection. We'll never be able to stop wildfires from starting, but the sooner we can detect burns and get suppression efforts under way, the better chance we'll have of keeping them from turning into massive conflagrations. Fortunately, early detection technologies are growing at a rapid pace. Wireless video surveillance technology, for example, is helping land managers located many miles away detect smoke in the air and dispatch the appropriate response. Infrared sensors—which can be used either on the ground, from airplanes, or from satellites—have the advantage of being able to detect heat from burns even through clouds or fog.

But the path ahead is not just a prompt for scientific and technological achievement. Beyond that, it's an opportunity for all of us to start believing again in the power of community and working in concert toward a common aim. Jean-Michel Cousteau wrote, "We have the capacity to design our own future, to take a lesson from living things around us and bring our values and actions in line with ecological necessity. But we must first realize that ecological and social and economic issues are all deeply intertwined. There can be no solution to one without a solution to the others."

Unquestionably, at this point we've gained sufficient understanding to allow us to act in ways that can make things better. We know that reducing fuel loads requires prescribed and managed burns, and here and there commercial thinning. Our wildfire risk models are good enough to be used as a solid platform for improving local and national policy decisions—from finding better ways to fund firefighting and wildfire science to encouraging smarter development and maintenance of the wildland-urban interface. Maybe we can also find ourselves making better choices to ease the human contribution to climate change: finding more efficient ways to heat and cool our buildings, growing the food we eat, reducing the footprint of the vehicles we drive, using the planet's oceans and forests in sustainable fashion. Maybe one day we'll even insist on leaders having the wisdom and courage to neither deny such challenges nor freeze in the face of them.

But first and foremost, there's simply this: fully accepting the fact that much of North America, especially the West, is being shaped by wildfire like never before. Wildfire is now, and will continue to be for many decades to come, a frequent companion—changing in ways both big and small how we, as well as tens of thousands of other species, experience life on Earth.

A FINAL WORD

AS I WRITE THIS IN MY OFFICE IN SOUTHWEST MONTANA, ON A warm day in mid-March, just outside the window is a sprawling grove of aspen trees. For most of the nearly thirty years I've lived here, especially in the first two decades, these trees would've been standing in generous blankets of snow. But today they rise from bare ground. The snowpack this year in this particular drainage, at least at this elevation, is just 8 percent of normal; indeed, the last major snowstorm was back at the end of December. And so far, temperatures for the year have been well above average.

With that in mind, as you might expect the conversations in town are already turning to the subject of wildfire. Most of us aren't overly pessimistic—at least not yet—because experience has taught us that if we have a wet June and July in the northern Rockies, fire danger will remain pretty low. On the other hand, rains like that aren't always so easy to come by.

And if that moisture doesn't come, before long you'll find us, like so many other people in communities all around the West, frequently casting our eyes up to any sky that has a smear of dark clouds—watching not so much for rain but for flashes of lightning. We'll be walking out of our houses into the chill of morning, all the while sniffing the air for smoke. And at some point, probably in late July or early August, you may well find us gathering up our precious belongings—photos and computers and musical instruments and family keepsakes—and loading them into cardboard boxes that we pile by the front door, all so we can load them quickly into the car should word come that a wildfire is coming, that it's time to evacuate.

Those of us who elect to keep living in the craggy, enchanted landscapes of the American West, so rich with big sky and bright mountains and hushed forests, are now faced with making some measure of peace with this new world, to learn how to dance to a tune increasingly driven by big burns. Some of us, myself included, hope things will change, that one day the climate may settle back into patterns more similar to those that have been more or less reliably in play for centuries. But if we're honest, this hope we share is likely to bear fruit only long after we're gone. For now, and probably for a great many decades to come, we'll be living in the middle of a thoroughly arresting yet increasingly daunting landscape. A turbulent and often overwhelming land of fire.

SOURCE NOTES

Living Fire

Statistics on suppression costs (1985–2015), historically significant wildland fires, and 1997–2014 large fires are from the National Interagency Fire Center, https://www.nifc.gov/fireInfo/fireInfo_statistics.html.

"As Wildfires Continue to Burn, New Maps [sic] Shows Expansion of Wildland-Urban Interface," USDA Office of Communications news release, September 10, 2015.

Whit Bronaugh, "North American Forests in the Age of Man," *American Forests*, Summer 2012.

Cynthia Carey and Michael A. Alexander, "Climate Change and Amphibian Declines: Is There a Link?" *Diversity and Distributions* 9 (March 2003): 111–121.

Climate Central, "The Age of Western Wildfires: Western Wildfires 2012," September 2012.

Ross Gorte, "The Rising Cost of Wildfire Protection," a research paper from Headwaters Economics, June 2013.

International Association of Wildland Fire, "WUI Fact Sheet," August 1, 2013.

Charles E. Kay, "Native Burning in Western North America: Implications for Hardwood Forest Management," in Daniel A. Yaussy, ed., *Proceedings: Workshop on Fire, People, and the Central Hardwoods Landscape* (Newtown Square, PA: U.S. Department of Agriculture, Forest Service, Northeastern Research Station, 2000).

Becky K. Kerns and Qinfeng Guo, "Climate Change and Invasive Plants in Forests and Rangelands," U.S. Department of Agriculture, Forest Service, Climate Change Resource Center, September 2012.

Rebecca Kessler, "Followup in Southern California: Decreased Birth Weight Following Prenatal Wildfire Smoke Exposure," *Environmental Health Perspectives* 120 (September 2012): A 362.

Oregon Health Authority, "Wildfire Smoke and Your Health," April 2014.

Laura Parker, "How Megafires Are Remaking American Forests," *National Geographic*, August 9, 2015.

Society for Conservation Biology Scientific Panel on Fire in Western U.S. Forests (Reed F. Noss, ed.; Jerry F. Franklin, William L. Baker, Tania Schoennagel, and Peter B. Moyle), *Ecology and Management of Fire-prone Forests of the Western United States* (Arlington, VA: Society for Conservation Biology, North American Section, August 2006).

Tom Tidwell, chief of the U.S. Forest Service, "Learning to Live with Fire," speech presented at the Large Wildland Fires Conference, Missoula, MT, May 19–23, 2014.

Kindling

Timothy Egan, *The Big Burn: Teddy Roosevelt and the Fire That Saved America* (New York: Houghton Mifflin Harcourt, 2009).

Edward Stahl, comment on the Big Burn from his undated personal notes (typescript), www.foresthistory.org/ASPNET/Policy/Fire/FamousFires/EdwardStahl.pdf.

The Forest History Society, "The 1910 Fires," www.foresthistory.org/ASPNET/Policy/Fire/FamousFires/1910Fires.aspx.

Robert E. Keane, Kevin C. Ryan, Tom T. Veblen, Craig D. Allen, Jesse Logan, and Brad Hawkes, "Cascading Effects of Fire Exclusion in Rocky Mountain Ecosystems: A Literature Review," U.S. Department of Agriculture, Forest Service, Rocky Mountain Research Station, May 2002.

Pacific Biodiversity Institute, "Fire Ecology," www.pacificbio.org/initiatives/fire/fire_ecology.html.

A. Park Williams, Richard Seager, John T. Abatzoglou, Benjamin I. Cook, Jason E. Smerdon, and Edward R. Cook, "Contribution of Anthropogenic Warming to California Drought During 2012–2014," *Geophysical Research Letters* 42 (August 2015): 6819–6828.

Combustion

Mark A. Finney, "FARSITE: Fire Area Simulator—Model Development and Evaluation," U.S. Department of Agriculture, Forest Service, Rocky Mountain Research Station, March 1998, revised February 2004.

Eunmo Koo, Patrick J. Pagni, David R. Weise, and John P. Woycheese, "Firebrands and Spotting Ignition in Large-scale Fires," *International Journal of Wildland Fire* 19 (2010): 818–843.

National Weather Service, National Oceanic and Atmospheric Administration, "How Lightning Is Created," www.srh.noaa.gov/srh/jetstream/lightning/lightning.html.

National Wildfire Coordinating Group, *Communicator's Guide for Wildland Fire Management: Fire Education, Prevention, and Mitigation Practices*, no date.

——, *Wildfire Origin and Cause Determination Handbook*, July 2014.

Rachel G. Schneider (USDA Forest Service) and Deborah Breedlove (Georgia Forestry Commission), *Fire Management Study Unit*, no date.

U.S. Department of the Interior, National Park Service, "The Fire Triangle," www.nps.gov/fire/wildland-fire/learning-center/fire-in-depth/fire-triangle.cfm.

Natalie S. Wagenbrenner, Jason N. Forthofer, Brian K. Lamb, Kyle S. Shannon, and Bret W. Butler, "Downscaling Surface Wind Predictions from Numerical Weather Prediction Models in Complex Terrain with WindNinja," *Atmospheric Chemistry and Physics* 16 (2016): 5229–5241.

Fighting Fire

Kelsi Bracmort, "Wildfire Protection in the Wildland-Urban Interface," Congressional Research Service, January 30, 2014.

National Wildfire Coordinating Group, *Wildland Fire Suppression Tactics Reference Guide*, April 1996.

U.S. Department of Agriculture, Forest Service, "History of Smokejumping," in *National Smokejumper Training Guide*, 2008.

——, *National Study of Tactical Aerial Resource Management to Support Initial Attack and Large Fire Suppression*, Final Committee Report, October 1998.

———, "Wildland Fire Shelter: History and Development of the New Generation Fire Shelter," no date.

U.S. Department of the Interior, National Park Service, "Incident Command System (ICS)," www.nps.gov/fire/wildland-fire/learning-center/fire-in-depth/incident-command-system.cfm.

Elizabeth Walatka, ed., *Yarnell Hill Fire: Serious Accident Investigation Report*, September 2013

Aftermath

Susan H. Cannon and Jerry DeGraff, "The Increasing Wildfire and Post-Fire Debris-Flow Threat in Western USA, and Implications for Consequences of Climate Change," in *Landslides: Disaster Risk Reduction* (World Landslide Forum, 2008).

Susan H. Cannon, Joseph E. Gartner, Raymond C. Wilson, James C. Bowers, and Jayme L. Laber, "Storm Rainfall Conditions for Floods and Debris Flows from Recently Burned Areas in Southwestern Colorado and Southern California," *Geomorphology* 96 (April 2008): 250–269.

J. Greenlee, ed., *The Ecological Implications of Fire in Greater Yellowstone: Proceedings of the Second Biennial Conference on the Greater Yellowstone Ecosystem* (Fairfield, WA: International Association of Wildland Fire, 1993).

Monica Turner, "Ecological Effects of the '88 Yellowstone Fires," *Yellowstone Science* 17 (2009): 24–30.

U.S. Department of the Interior, National Park Service, "Yellowstone: Ecological Consequences of Fire," www.google.com/search?q=after+the+Yellowstone+fires&ie=utf-8&oe=utf-8.

U.S. Geological Survey, "Fire Ecology," www.werc.usgs.gov/ResearchTopicPage.aspx?id=6.

———, "Wildland Fire Research, www2.usgs.gov/ecosystems/environments/fireecology.html.

Risk Reduction

Headwaters Economics, "Reducing Wildfire Risks to Communities: Solutions for Controlling the Pace, Scale, and Pattern of Future Development in the Wildland-Urban Interface," 2014.

New Mexico Environment Department, "Wildfire Impacts on Surface Water Quality," www.env.nm.gov/swqb/Wildfire/.

Northwest Fire Science Consortium, "Communication Under Fire: Communication Efficacy During Wildfire Incidents," Research Brief 7, 2016.

Joel Rubin, Andrew Blankstein and Scott Gold, "A searingly familiar scene," *Los Angeles Times*, 22 October 2007.

Christopher Topik, "Wildfires Burn Science Capacity," *Science* 349 (18 September 2015): 1263.

U.S. Department of Agriculture, Forest Service, "Protecting Residences from Wildfires," www.fs.fed.us/psw/publications/documents/gtr-050/nature.html.

Wildland Fire Leadership Council, *The National Strategy: The Final Phase in the Development of the National Cohesive Wildland Fire Management Strategy*, April 2014.

Future Fire

Lisa-Natalie Anjozian, "ArcFuels: Integrating Wildfire Models and Risk Analysis into Landscape Fuels Management," *Fire Science Brief* 43 (February 2009): 1–5.

Nathanael Massey, "As Wildfires Rage in U.S. West, Scientists Predict Worse Blazes in Future," *Scientific American*, June 14, 2012.

Simon Wang, "How Might El Niño Affect Wildfires in California?" Climate.gov blog post, August 27, 2014.

Amanda M. West, Sunil Kumar, and Catherine S. Jarnevich, "Regional Modeling of Large Wildfires under Current and Potential Future Climates in Colorado and Wyoming, USA," *Climatic Change* 134 (2016): 565–577.

FURTHER READING

Timothy Egan, *The Big Burn: Teddy Roosevelt and the Fire That Saved America* (New York: Houghton Mifflin Harcourt, 2009).

John N. Maclean, *Fire and Ashes: On the Front Lines of American Wildfire* (New York: Holt, 2003).

Norman Maclean, *Young Men and Fire* (University of Chicago Press, 1992).

Brian Mockenhaupt, "Fire on the Mountain," *Atlantic*, June 2014.

Laura Parker, "How Megafires Are Remaking American Forests," *National Geographic*, August 9, 2015.

Stephen J. Pyne, *America's Fires: A Historical Context for Policy and Practice* (Forest History Society, reprint edition, 2010).

———. *Fire: A Brief History* (University of Washington Press, 2001).

———. *Fire in America: A Cultural History of Wildland and Rural Fire* (University of Washington Press, 1997).

Paul Tullis, "Into the Wildfire," *New York Times Magazine*, September 19, 2013.

ACKNOWLEDGMENTS

SINCERE THANKS TO ALL THE SCIENTISTS, FIREFIGHTERS, AND LAND managers who are devoting their lives to the understanding and management of wildfire in the American West. This book would not have been possible without their critical insights and generous support. A special thanks to the brilliant Jon Trapp and to his extraordinarily devoted colleagues at the Red Lodge, Montana, Fire Rescue Department.

PHOTO AND ILLUSTRATION CREDITS

Illustration by Anna Eshelman based on original by Christophe Dang Ngoc Chan SVG conversion JBarta, Wikimedia/Used under a Creative Commons Attribution-ShareAlike 3.0 Unported License, page 78

Dr. Richard Miller, Oregon State University, page 162 top

Flickr/Justin Peterson Photography, page 169

Flickr/Rachael Botkin (Newhart), page 19 bottom

Gary Ferguson, page 173

Kari Greer, pages 2, 28, 58, 68, 69, 73, 105 right, 109, 111, 117, 120, 125, 126, 146 right–147, 177

Kari Greer/NIFC, page 159

Shutterstock/Pi-Lens, page 132 left

State Farm Insurance, page 32 top

Flickr

Used under a Creative Commons Attribution-ShareAlike 2.0 Generic License

U.S. Forest Service, Coconino National Forest, photo by Brienne Magee, page 188

U.S. Forest Service, Gila National Forest, photo by Kari Greer, pages 77, 114–115

U.S. Forest Service, Southwestern Region, Kaibab National Forest, pages 103, 105 left

Tim Gage, page 53 bottom

Torroid, page 71

Used under a Creative Commons Attribution 2.0 Generic License

Anthony Quintano, page 65

Justin Fox, courtesy of Edward Vielmetti and Jason Fox, page 31

NASA, photo by Jeff Schmaltz, page 178

National Park Service, Sarah Stio, page 155

Ron Reiring, page 170

U.S. Air National Guard, photo by Master Sergeant Julie Avey, page 25

U.S. Army, photo by Staff Sergeant Malcolm McClendon, pages 123, 124

U.S. Bureau of Land Management, pages 92, 128

U.S. Bureau of Land Management, photo by Jeffrey McEnroe, pages 120 right–121

U.S. Department of Agriculture, photo by Lance Cheung, pages 112 right–113, 144 bottom

U.S. Forest Service, pages 144 top, 146, 168

U.S. Forest Service, photo by Roy Bridgeman, page 140

U.S. Forest Service, photo by Kari Greer, page 107

U.S. Forest Service, photo by Mike McMillan, page 80

U.S. Forest Service, photo by Doc Searls, page 179

U.S. Forest Service, Southwestern Region, Kaibab National Forest, photo by Dyan Bone, page 82

Zach Dischner, pages 34–35

Used under a Creative Commons Attribution-NoDerivs 2.0 Generic License

State of Washington, page 127

Pexels

Used Under a Creative Commons Zero License

Tim Mossholder, pages 22–23

Public Domain Images

Federal Emergency Management Agency, photos by Andrea Booher, pages 21, 95, 112

Federal Emergency Management Agency, photo by Michael Mancino, page 19 top

Federal Geographic Data Committee (FGDC), 2008 Annual Report / Kevin Hyd, WFDSS, page 87

LANCE/EOSDIS MODIS Rapid Response Team, GSFC, NASA, pages 36–37

National Interagency Fire Center, page 14

National Park Service, page 138

National Park Service, photo by Mike Lewelling, pages 48–49

National Park Service, photo by Jim Peaco, pages 131, 132 right, 134 left, 150

National Park Service, Yellowstone National Park, photo by Jeff Henry, pages 84, 99, 149

National Photo Company, pages 42–43

Natural Resources Conservation Service, photo by Jeff Vanuga, page 148

NOAA Geophysical Fluid Dynamics Laboratory, pages 184, 185

NOAA Photo Library, page 181

Oregon Air National Guard, photo by James Haseltine, pages 82–83

Oregon Department of Forestry, photo by Marvin Vetter, page 85

U.S. Air Force, photo by Senior Master Sergeant Dennis W. Goff, pages 26–27

U.S. Air Force, Master Sergeant Christopher DeWitt, page

U.S. Fish and Wildlife Service, photo by Scott Swanson, page 161

U.S. Forest Service, photo by Eric Knapp, page 62

U.S. Forest Service, photo by John McColgan, edited by Fir0002, page 137

U.S. Forest Service, Incident Information System, page 162 bottom

U.S. Geological Survey, photo by Don Becker, page 55
U.S. Geological Survey, photo by Michael Lewis, page 142
U.S. Geological Survey, photo by Nathan Stephenson, page 187
U.S. Marine Corps, page 32 bottom
U.S. National Archives and Records Administration, page 45 right
U.S. National Guard, Oregon Military Department, page 122

Wikimedia Commons

Used under a Creative Commons Attribution-ShareAlike 3.0 Unported License

Brambleshire, page 41
Dcrjsr, page 57
Famartin, page 135
GeorgeLouis, page 53 top
Twelvizm, pages 50–51
Walter Siegmund, page 133
Mav, page 44–45

Used under a Creative Commons Attribution 3.0 United States License

Chris Schnepf, University of Idaho, page 17

Used under a Creative Commons Attribution 3.0 Unported License

CSIRO/Robert Kerton, page 134 right

Used under a Creative Commons Attribution-ShareAlike 2.0 Generic License

Ben Brooks, pages 10–11

INDEX